Culture as Renewable Oil

This book unpacks the links between oil energy, state power, urban space and culture, by looking at the Petro-Socialist Venezuelan oil state. It challenges the disciplinary compartmentalisation of the analysis of the material and cultural effects of oil to demonstrate that within the Petrostate, Territory, Bureaucratic Power and Culture become indivisible. To this end, it examines how oil is a cultural resource, in addition to a natural resource, implying therefore that struggles over culture implicate oil, and struggles over oil implicate culture. This book develops a story about Venezuela as an oil state and the way it deploys its policies to instrumentalise culture and urban space by examining the way Petro-Socialism manifests in space, how it is imagined in speeches and how it is discursively constructed in adverts. The discussion reveals how a particular culture is privileged by the Venezuela state-owned oil company and its social and cultural branch. The book explores to what effect the state-owned oil company constructs a parallel notion of culture that becomes inextricable from land, akin to a mineral deposit, and tightly controlled by the Petrostate.

The book will appeal to researchers who are interested in Resource Management, Environmental Studies, Cultural Studies and Political Geography.

Penélope Plaza Azuaje is a Venezuelan architect, researcher and urban artivist. She researches the entanglements between oil, politics, culture and urban space, with a particular interest in urban artivism, and contemporary Venezuelan petro-politics. She holds a PhD in Cultural Policy and Management from City University of London and an MPhil in Latin American Studies from the University of Cambridge. Currently she is a Lecturer in Architecture at the School of Architecture, University of Reading.

Routledge Research in Place, Space and Politics

Series Editor: Professor Clive Barnett

University of Exeter, UK

This series offers a forum for original and innovative research that explores the changing geographies of political life. The series engages with a series of key debates about innovative political forms and addresses key concepts of political analysis such as scale, territory and public space. It brings into focus emerging interdisciplinary conversations about the spaces through which power is exercised, legitimised and contested. Titles within the series range from empirical investigations to theoretical engagements and their authors comprise scholars working in overlapping fields including political geography, political theory, development studies, political sociology, international relations and urban politics.

The Politics of Settler Colonial Spaces
Forging Indigenous Places in Intertwined Worlds
Edited by Nicole Gombay and Marcela Palomino-Schalscha

Direction and Socio-spatial Theory
A Political Economy of Oriented Practice
Matthew G. Hannah

Postsecular Geographies
Re-envisioning Politics, Subjectivity and Ethics
Paul Cloke, Christopher Baker, Callum Sutherland and Andrew Williams

Culture as Renewable Oil
How Territory, Bureaucratic Power and Culture Coalesce in the Venezuelan Petrostate
Penélope Plaza Azuaje

Migration in Performance
Crossing the Colonial Present
Caleb Johnston and Geraldine Pratt

For more information about this series, please visit: www.routledge.com/series/PSP

Culture as Renewable Oil

How Territory, Bureaucratic Power and Culture Coalesce in the Venezuelan Petrostate

Penélope Plaza Azuaje

Routledge
Taylor & Francis Group

LONDON AND NEW YORK

First published 2019
by Routledge
2 Park Square, Milton Park, Abingdon, Oxon OX14 4RN

and by Routledge
52 Vanderbilt Avenue, New York, NY 10017

First issued in paperback 2020

Routledge is an imprint of the Taylor & Francis Group, an informa business

British Library Cataloguing-in-Publication Data
A catalogue record for this book is available from the British Library

Library of Congress Cataloging-in-Publication Data
A catalog record has been requested for this book

ISBN 13: 978-0-367-58205-0 (pbk)
ISBN 13: 978-1-138-57377-2 (hbk)

Typeset in Times New Roman
by Wearset Ltd, Boldon, Tyne and Wear

Contents

Acknowledgements   vii

Introduction   1

The making of a modern oil nation 2
Interfaces of State Space, Bureaucratic Power and Culture as a
    Resource 7
Chapter outline 9

1   **Entanglements of oil, modernity, state and culture in**
    Venezuela   14

Oil and the illusion of modernity 15
Fleeting mirages of oil wealth: from Pérez's Great Venezuela to
    Chávez's Petro-Socialism 30
Conclusion 40

2   **Oil in the intersection between Territory, Bureaucratic**
    Power and Culture as a Resource   46

State Space as territory 46
Bureaucratic Power 54
Culture as a Resource in and within the Petrostate 65
Conclusion 75

3   **Territory effect, the New Geometry of Power and the**
    construction of a Petro-Socialist State Space   78

The New Geometry of Power and the construction of the
    Petro-Socialist State Space 79
PDVSA's Oil Social District as a parallel Petro-Socialist State
    Space 88
Conclusion 92

**4   Bureaucratic Power, performative speech and oil policy:
     'sow the oil' to 'harvest culture'**   96

The performative power of discourse 98
Hugo Chávez's discursive construction of Petro-Socialism 101
*Rafael Ramírez and the new PDVSA as agents of
     Petro-Socialism 106*
*PDVSA La Estancia, an instrument of the Sowing Oil Plan that
     'harvests culture' 109*
Conclusion 115

**5   Giant oil workers and the expediency of Culture as
     Renewable Oil**   121

The expediency of culture as a mineral resource 122
*Profile of the campaign 'we transform oil into a renewable
     resource for you' 123*
*Visual semiotics of oil: giant oil workers and the myth of Culture
     as Renewable Oil 124*
Giant oil workers and the myth of Culture as Renewable Oil 129
Conclusion 135

Conclusion: the untenable utopia of oil   138

Oil as cultural culprit 142

Index   145

Acknowledgements

This book on Venezuela was very challenging to write; it wouldn't have been possible without the encouragement and support of a small but significant group of people. I would like to thank my parents and my brothers for their support, resilience and for maintaining a sense of humour despite the difficulties they face in Venezuela and the distance that currently separates us. I want to thank Dr Dave O'Brien, my PhD supervisor at City University of London, and Professor Flora Samuel, Professor of Architecture at University of Reading, for encouraging me to progress my doctoral research into this monograph. I am very grateful to Dr Tomás Straka, Professor at Universidad Católica Andrés Bello and member of Academia Nacional de la Historia of Venezuela, for reading sections of the book, his work and input did much to shape the book, and particularly Chapter 1. And last but not least, I am forever indebted to Oliver Froome-Lewis, whose unwavering support and enthusiasm was fundamental in preserving my sanity and in making this book a reality.

Introduction

It's July 2014. On a weekly routine trip to the nearest PDV petrol station in Caracas, while waiting for the tank to fill (Venezuela has the cheapest fuel in the world) I looked up to the fuel dispenser and its empty shelves. Instead of the usual adverts for PDV lubricants or motor oil, it featured the nineteenth century white filigree gazebo of El Calvario park in Caracas; looming over it appeared a Gulliver-scale version of an oil worker wearing red coloured gear. I got out of my car to take a photograph and noticed that the dispenser on the opposite lane also had a similar advert with another giant oil worker grazing the multicolour ceramic mural that covers Libertador Avenue in Caracas.

The presence of the giant oil workers in the images signalled that something different was at play in the manner in which the state-owned oil company Petróleos de Venezuela Sociedad Anónima (PDVSA) has been extending its dominion over the city and its cultural symbols. Within contemporary scholarly work on the politics of culture of Hugo Chavez's Bolivarian revolution (Kozak Rovero, 2014, 2015; Silva-Ferrer, 2014), little attention has been paid to PDVSA's recent interventions in the city, which I regard as a clear sign of the increasing power Hugo Chávez had granted to the state-owned oil company. Amid the myriad of recent publications on the cultural representations of oil capitalism in pop culture, literature and the visual arts (Barrett and Worden, 2014; LeMenager, 2014; Lord, 2014), far less attention has been paid to the spatial dimension of the material cultural effects of oil, both as a mineral and as a flow of energy, political power and wealth. Furthermore, recent cultural studies of oil have been predominantly focused on European and North American oil producing countries, with little focus on the Global South or OPEC countries more specifically. This book sets out to challenge the disciplinary compartmentalisation of the analysis of the material and cultural effects of oil. Tim Mitchell's *Carbon Democracy* (2011) marks the point of departure of this book's approach to look beyond the attention confined to the allocation of oil money to examine the processes through which oil flows are converted into political and cultural power (Mitchell, 2011, pp. 5–6). The particular case of the Venezuelan Petrostate in the era of Petro-Socialism serves to develop a reconsideration of the premises behind the cultural analyses of oil. Historically, the formation of modern statecraft and society in Venezuela is inextricable from the oil industry;

the influence of oil cannot be confined to a set of tropes or circumscribed to punctual interventions in the public sphere.

Hugo Chávez shifted the relationship between PDVSA and the state by making the state-owned oil company subservient to his centralist political project of Petro-Socialism, further coalescing oil, territory, state and culture.

This book examines the discursive and spatial dimensions of the entanglement between oil, territory, Bureaucratic Power and culture in the contemporary Venezuelan Petrostate. To develop these themes, this introductory chapter is divided into three parts. The first part sketches the historical context of this study, situating it within debates around the pervasive presence of oil in the formation of modern statecraft in Venezuela and the shift in the relationship between oil, modernity and statecraft brought by Hugo Chávez's Petro-Socialism. The second part presents the theoretical premises that inform this book and identifies the key themes that will be developed throughout, how in a Petrostate, oil traverses territory, Bureaucratic Power and culture. Finally, the third part presents the chapter outlines, providing an introduction to the discussions and main arguments developed in this book.

The making of a modern oil nation

It wasn't until the rise of the oil industry in the early twentieth century that Venezuela acquired the economic and political resources to develop modern statecraft with a centralised bureaucracy. But due to the strong legacy of Spanish colonial rule, decisions were founded on the traditions inherited from the colony, as the emerging nation declared itself the heir of the property rights of the Spanish Crown over vacant lands and ownership over all mines (Pérez Schael, 1993, p. 39). The property rights derived from the principle that what belonged to no one belonged to the King, so after gaining independence from the Spanish Crown, the new republic substituted the King (Pérez Schael, 1993, p. 39). The wealth extracted directly from the subsoil as rent became an affirmation of national sovereignty, the rent sanctioned the recognition of the nation's authority as analogous to the King's. Venezuela did not become a rentier state with the rise of the oil industry, it was born a rentier landlord state from the moment it became a modern republic and with the exploitation of oil the country inaugurated its modern history as a Petrostate. A Petrostate is a particular form of the rentier state, in which the majority of the state's revenue comes from abroad through oil exports. The concept of the 'rentier state' was coined by Iranian economist Hossein Mahdavy (Beblawi and Luciani, 1987, p. 9) to refer to states whose main source of income comes from external resources, 'one whose capacity to create consensus and enforce collective decisions rested largely on the fate of the international oil market' (Karl, 1997, p. 91). Rentier states can be traced back to the seventeenth century Spanish Empire and its exploitation of the vast mineral resources found in the Americas.

The sovereignty of the state was built around the notion of property thus annulling the mineral materiality of oil, reducing it to the fetish of rent money that flowed from the subsoil directly to the state's coffers; therefore oil mattered as money and not as a complex technological new reality (Pérez Schael, 1993, p. 94). Early twentieth century Venezuela was still a rural society with a predominantly agricultural economy dependent on exports of cocoa and coffee vulnerable to fluctuations of the international market, and deeply engulfed by political turmoil, debt and military unrest (Mommer, 1994, p. 27; Baptista, 1997, p. 131). Thus, the country lacked the resources and the capacity to exploit and produce oil products, the only option for the state to secure revenue came in the form of concessions and royalties (Harwich Vallenilla, 1984). For the Venezuelan state the only matter to resolve was where, how and to who distribute oil rent money, not how to produce it.

Landowners, traditional elites and the intellectual class attributed an ephemeral quality to oil wealth because unlike agriculture, wealth relied on rent money and not on produce (Pérez Schael, 1993, p. 95). This apprehension and rejection was grounded on the invisibility of crude oil; it is a material entity hidden in the subsoil, its potential yield not as visibly quantifiable as land and hectares of crops on the surface. When foreign oil companies began to establish in Venezuela in the early twentieth century, the material effects of oil wealth were not felt immediately on the areas surrounding oil drills and refineries, they materialised first in remote oil camps and fenced residential quarters built by foreign oil corporations, enclaves of modernity frequently surrounded by poverty belts (González Casas and Marín Castañeda, 2003, p. 381). The iron fences built to isolate the oil camps did not stop poor neighbouring communities peeping into the modernity of foreign capital: technology, urban planning, architecture, corporate culture and lifestyle (González Casas and Marín Castañeda, 2003, pp. 381–382). The belief shared by many intellectuals throughout the twentieth century that oil wealth had become a colonising and demonic force (Pérez Schael, 1993, p. 9) is clearly expressed in the influential work of Marxist anthropologist and former oil camp dweller Rodolfo Quintero, who wrote in 1968 the influential essay titled *The Culture of Oil: Essay on the Life Styles of Social Groups in Venezuela* (2011). Here Quintero defined the 'culture of oil' as a foreign force of conquest with its own technology, instruments, inventions, equipment and non-material devices such as language, art and science that decimates local and indigenous cultures (2011, pp. 19–20), sustained by the exploitation of national oil wealth by way of monopolistic foreign companies. Quintero unequivocally demonised the United States, oil wealth, rapid urbanisation and bureaucratic and technological apparatuses as predators and destroyers of national culture, dividing the history of Venezuela into an idyllic pre-oil era and a culture of oil era that brought the oil camp and the oil city (2011, p. 25).

Novelist, essayist and politician Arturo Uslar Pietri coined the slogan 'to sow the oil' in a newspaper article published in 1936 (1936) using farming language as a didactic trope – oil becoming akin to a rare 'seed' – to propose that the oil windfall should be invested in development and modernisation of the agricultural economy, making a direct reference to the land where the 'seed' is sown and riches are harvested from, and not to oil as an ephemeral source of wealth and dependency (Straka, 2016, p. 139). Uslar Pietri was an advocate of taking advantage of the knowledge, technology and financial power of foreign oil corporations; he saw great benefits in keeping the country open to foreign capital, using it to invest in economic and social development, under the leadership of an illustrated elite, the owners of the 'seed' (Uzcátegui, 2010, pp. 37–38; Urbaneja, 2013, pp. 81–89). But by the late 1940s Uslar Pietri identified an emerging 'feigned nation' (Uslar Pietri, 2001; Straka, 2016, p. 140) with a parapet of modernity built upon a transient oil wealth that once exhausted would lay bare the 'real nation', still backwards. Nonetheless, 'to sow the oil' became a guiding principle of political and economic policy of subsequent governments (Coronil, 1997, p. 134). Uslar Pietri's slogan 'to sow the oil' is at the centre of enduring conflicting views around oil. Behind the belief that oil can be 'sown' there is a lingering nostalgia about a lost idyllic agrarian past that created a tension of simultaneous embrace and demonisation of oil.

Fernando Coronil's seminal study *The Magical State: Nature, Money and Modernity in Venezuela* (1997) argues that the Venezuelan Petrostate exercised its monopoly over the oil rent dramaturgically, enacting collective fantasies of progress by way of spectacular projects of development and infrastructure to seize its subjects through the power of marvel rather than with the power of reason: 'the state seizes its subjects by inducing a condition or state being receptive to its illusions – a magical state' (1997, p. 5). In a country where the state had historically been very weak, the expansion of the oil industry promoted the concentration of power in the presidency, the embodiment of the 'magical' powers of oil; the Magical State is personified as a magnanimous sorcerer in the figure of the president. Venezuela's identity as a nation is deeply entangled with oil; as the Petrostate engaged with the oil industry Venezuelan society learned to see itself as an oil nation with the state as the single representative of a population unified by oil (Coronil, 1997, p. 84). The Venezuelan Petrostate came to be viewed as an enormous distributive apparatus of oil rent money, increasingly hollowed out by a breach between authority and territory, modernity and modernisation. Moreover, the oil industry in Venezuela exercised a pervasive influence on the formation of political and social values promoting and influencing the emergence of a political and social order based on the entrepreneurial corporate model of the oil industry (Tinker Salas, 2014, pp. 12–13). Oil wealth suddenly made possible lavish and monumental works of infrastructure (Coronil, 1997, p. 76) for a country that had been in chronic debt and lacked basic infrastructure such as a national road network and systems of communications. And as oil wealth increased, so did the capacity of the Petrostate to construct itself as a national institution by expanding the range of its dominion over society with material illusions of progress through massive works of infrastructure and a vertiginous process of urbanisation achieved in just a few decades.

Venezuela gave absolute freedom to foreign capital transactions, but this began to change in the 1970s. The oil booms of 1973 and 1979 produced significant increases in oil revenues. This windfall prompted the then president, Carlos Andrés Pérez, to launch the ambitious development plan The Great

Venezuela, promising that the increased financial power of the state would allow Venezuela to 'catch up' and become a developed country in just a few years. He nationalised the oil industry in 1975 and in 1976 created the state-owned oil company Petróleos de Venezuela Sociedad Anónima (PDVSA) (Darwich, 2008, p. 50), carried out an ambitious programme of infrastructure and project of state reform. But this oil windfall was soon followed by a dramatic plunge in oil prices. The decade that followed The Great Venezuela was one of gradual economic and social decline (López-Maya, 2006, p. 21; Urbaneja, 2013, p. 279), it suffered its final blow on 18 February 1983, when President Luis Herrera Campins devalued the national currency in the aftermath of a dramatic dip in oil prices in 1982, opening a cycle of economic stagnation, high inflation, increase of foreign debt, with the resulting deterioration of quality of life for large sectors of the population (Salamanca, 1994, p. 11; López-Maya, 2006, pp. 22–23).

With the steady decline in oil prices of the 1980s and 1990s, 'catching up' had become more elusive. Carlos Andrés Pérez was elected for a second presidency in 1988, under the illusion that he alone could summon a 'magical state' to revive the opulent days of The Great Venezuela (Atehortúa Cruz and Rojas Rivera, 2005, p. 264). But in 1989, with barely a month in office, he announced an IMF-backed programme of macroeconomic adjustments which most notably included a 100 per cent increase in the price of oil (López-Maya, 2003, p. 120). A country-wide popular revolt known as the *Caracazo* intensified the economic and social crisis. A political crisis unfolded in February 1992, when a small group of the army, with the support of leftist civilian groups, staged a failed coup d'état led by Lieutenant Coronel Hugo Chávez Frías (Coronil, 2000, p. 37; López-Maya, 2003, p. 129); a second failed coup d'état took place in November. Pérez's presidency survived the coups but it did not survive the deterioration of his political leadership; he was impeached and sentenced to house arrest in 1993 (Salamanca, 1994, p. 12).

A Petrostate that was navigating the tortuous path of consolidating a political consensus to modernise the country and develop an efficient bureaucratic apparatus modelled on oil overabundance fatally collided with the fall in oil prices, a rent-seeking political class and widespread corruption which aggravated the gradual collapse of institutional stability and social welfare. It would be inaccurate to conclude that the Petrostate failed to 'sow the oil'; throughout the democratic era of the Pact of Puntofijo the oil rent was invested in modernisation, infrastructure and industrialisation, but it was undermined by a deficient state apparatus and a rent-seeking political class unwilling to carry out necessary structural reforms. The exhaustion of the Pact of Puntofijo became the backdrop of Hugo Chávez presidential election by a landslide in 1998 on an anti-establishment political platform outlined as an alternative to neoliberalism.

The three consecutive presidencies of Hugo Chávez (1999–2001, 2002–2007, 2007–2013) are characterised as a 'new debut of the Magical State' (López-Maya, 2007; Coronil, 2011), there are close similarities between Chávez's government and the first presidency of Carlos Andrés Pérez in the centralisation of power and the use of the oil windfall to completely reform the state (López-Maya, 2007).

A new oil boom between 2003 and 2008 surpassed that of the 1970s (Corrales and Penfold, 2011, pp. 55–57), that translated into an increase in public spending, the politicisation of PDVSA and a radicalisation of Chávez's political project then onwards. In the midst of this unprecedented rise in oil prices Chávez launched the Plan Siembra Petrolera (Sowing Oil Plan), a 25-year national plan and oil policy that formed the foundation to lay the groundwork for the transition towards the Socialist State. Aware of the diminished capacity of the public sector, Chávez believed that 'an oil company would succeed where government ministries might not' (Maass, 2009, p. 215). He altered the established institutional channels for the flow of oil rent from the state-owned oil company to the state. PDVSA was put in charge of new government programs, effectively transforming the oil company into the 'engine of revolutionary change' (Maass, 2009, pp. 202, 215), a direct life-line between PDVSA and public spending. He laid out his ambitions to transform Venezuela into a 'world energy power' formulated under the mirage that the oil windfall would be everlasting. He revived the use of the slogan 'to sow the oil' to frame the ambitions of his regime. As the new owner of the 'seed' (oil) he vindicated oil wealth for collective benefit, as had governments in the past (Urbaneja, 2013, pp. 81–89) with the difference that his Bolivarian revolution was set to achieve what previous governments could not: the 'harvest' of oil. Thus, territory was fundamental for Chávez's political project, although in practice the bureaucratic structures of the Socialist State had to coexist uncomfortably with the structures inherited from previous governments that it was meant to substitute.

Chávez declared his third presidential term (2007–2013) the dawn of a new era with the expansion of the Bolivarian revolution towards Socialism, the path for transcending capitalism. He assured that his socialist project was unique, that it was 'different to the Scientific Socialism that Karl Marx had originally envisioned' because he was building a Bolivarian, Venezuelan, oil-based *socialismo petrolero*, in other words, Petro-Socialism. Petro-Socialism broadly defines Hugo Chávez's political and economic project, in which the oil rent is funnelled into the construction of the Socialist State. Broadly speaking, Petro-Socialism is focused on using oil revenues to fund the transition towards a Socialist State and a new socialist society. Petro-Socialism is a peculiarly extreme form of oil rentierism. Underpinned by a steady rise in oil prices, the era of Petro-Socialism promised historically neglected social sectors that they would finally enjoy enduring prosperity provided by oil. The death of Hugo Chávez in March 2013 left the transition towards the Socialist State orphaned of its leader and mastermind. By then, Venezuela had become even more dependent on oil revenue than before.

Beyond a dramaturgical exercise of the monopoly over the oil rent, the close control over PDVSA enabled Chávez to summon all the bureaucratic powers of the state in his persona. But as will be made clear throughout this book, by transferring many of the bureaucratic powers of the state to PDVSA, Chávez paved the way for the state-owned oil company to exercise power as a parallel state and develop a discursive narrative that deploys the 'magical' power of marvel of the New Magical State confined to the realm of oil around a dual narrative of 'sowing oil' and 'harvesting oil' where culture becomes akin to 'renewable oil'.

Interfaces of State Space, Bureaucratic Power and Culture as a Resource

Three interlocked theoretical premises guide this book: State Space, Bureaucratic Power and Culture as a Resource. The first premise draws on Brenner and Elden's reading of Lefebvre as a theorist of State Space as territory. State Space is understood as land and as a political form of space which is historically specific, produced by and associated with the modern state, understandable 'only through its relation to the state and processes of statecraft'; accordingly, there can be no state without territory and no territory without a state (Brenner and Elden, 2009b, pp. 362–363). Brenner and Elden's reading of Lefebvre will be useful for this book as it provides a way to go beyond simplistic perceptions of territory by understanding that any State Space, and by extension, any 'territorially configured social space' is the consequence of specific historical forms of economic and political interventions of the state. This book engages critically with a diverse mix of documents and topics, it utilises David Harvey's (2006, pp. 281–284) matrix of categories of space as a taxonomy to locate the spatio-temporal category of each document (defined and described in detail in Chapter 2) in order to disentangle the spatial and discursive mechanisms that constituted the spatial policies deployed under Petro-Socialism.

The second premise is Bureaucratic Power. Bob Jessop posits that the state does not exercise power as the power of the state is 'always conditional and relational' it is defined as an institutional ensemble (Jessop, 1990, p. 367). Similarly, Nikolas Rose and Peter Miller (2008, p. 10) coincide with Jessop in arguing that the state does not and cannot exercise power, it can only do so through the complex network of organisations, institutions and apparatuses that compose it (2008, pp. 55–56). By the same token, Tony Bennett and Patrick Joyce affirm that the state 'rather than a site from which this form of power originates or at which it terminates' is the site where Bureaucratic Power congregates (Joyce and Bennett, 2010, p. 2). Hence, when referring to the power of the state it is more accurate to talk about Bureaucratic Power instead of state power. This book adopts Bennett and Joyce's perspective of the state as the site where Bureaucratic Power congregates to explore the contradictory process of transition towards the Socialist State, as it entailed the concurrent fragmentation of the existing institutional apparatus and the centralisation of Bureaucratic Power in the figure of President Hugo Chávez. The adoption of the Bureaucratic Power perspective also allows this book to integrate the idea of the state as an 'institution of territorial governance with vast powers over the material wellbeing of its people' (Mukerji, 2010, p. 82) considering that the modern state is the only agent with the capacity to manage territory on a large scale (Brenner and Elden, 2009a, p. 20). This book explores Bureaucratic Power as it derives from, and is subject to, the dominion over State Space as territory, a crucial notion in a Petrostate as its political and economic power originates from the ownership of the subsoil and the monopoly over the oil rent extracted from it.

The third premise constructs the notion of 'Culture as Renewable Oil', drawing on George Yúdice's expediency of Culture as a Resource. Yúdice's proposition is that culture has acquired to an extent the same status as natural resources as it is close to impossible to find public statements that do not instrumentalise art and culture, whether to improve social conditions or to foster economic growth (Yúdice, 2003, pp. 10–11). Through an exploration of the relationship between culture, management and power (McGuigan, 2003; Bauman, 2004; O'Brien, 2014) this book engages in particular with the social and cultural arm of PDVSA, PDVSA La Estancia, use of farming language and discursive fabrications to coalesce culture and oil ('PDVSA La Estancia is oil that harvests culture') to argue that for the state-owned oil company it is close to impossible not to turn to culture as a mineral resource, in which culture becomes akin to an implausible 'renewable oil'. Given that in practice, cultural policy is the bureaucratic medium for the instrumentalisation of Culture as a Resource (Miller and Yúdice, 2002, p. 1), this book also engages with Jeremy Ahearne's category of implicit cultural policy (Ahearne, 2009, p. 141) to demonstrate the use of the Organic Law of Hydrocarbons as a parallel instrument of territorial and cultural policy.

The three theoretical premises described above coalesce into a conceptual lens through the substantive chapters of this book, transcending the pitfalls of a compartmentalised analysis of the spatial and cultural dimension of oil, to demonstrate that within the Petrostate, oil inevitably intersects and interweaves State Space, Bureaucratic Power and Culture as a Resource. State Space as territory condenses the notion of land and the political space of statecraft; land being crucial to a Petrostate since the subsoil contains the deposits of crude oil that forms the basis of its financial and political power. The Bureaucratic Power of the Petrostate and its institutional apparatus relies and depends on the oil wealth extracted from the subsoil. Therefore, the manner in which the petrostate conceives and manages Culture as a Resource is framed within an oil rentierist logic, where culture is tantamount to oil as a resource inextricable from State Space.

In this regard, the discussion developed in this book contributes to current debates and recent scholarly work on the cultural dimension of oil, particularly within the emerging field of Energy Humanities. Thus, by building on the relationship between territory, Bureaucratic Power and culture, this particular tripartite theoretical lens provides the ideal framework to scrutinise how they function in the particular context of the contemporary Venezuelan Petrostate by addressing the relationship between the Petrostate, oil rentierism, statecraft and culture in Petro-Socialism. Looking at this relationship through this lens encourages the advancement of a new way of understanding the spatial and cultural dimensions of oil, and how a certain form of understanding culture is privileged by the national oil industry and to what effect it constructs a parallel notion of territorial and cultural policy making.

In summary, this book is concerned with investigating the discursive and institutional mechanisms that enabled the state-owned oil company to constitute a parallel State Space to extend its dominance over non-oil field spaces such as the city of Caracas, to effectively reframe the city as an oil field by discursively construing a notion of 'Culture as Renewable Oil' that ties culture to the land, where the 'sowing' of oil can 'harvest' culture. It is also concerned with the intrinsic contradictions within the model of Petro-Socialism that inform the paradoxical discursive notion of 'renewable oil' as an illusion of the New Magical State.

Chapter outline

Chapter 1 develops a historical account of the Venezuelan nation state in the context of postcolonial state formation in Latin America. It then develops a discussion on the emergence of the modern Venezuelan Petrostate coeval with the arrival of the oil industry and corporate practices of foreign oil corporations to illustrate how the Petrostate approached oil predominantly as rent money and not as a modern technological reality, which marked the emergence of what Fernando Coronil has termed the Magical State. The chapter builds on Coronil to characterise Hugo Chávez as the embodiment of the New Magical State and PDVSA as the engine of his revolution. Finally, the chapter provides a discussion on the historical context of the intersections between oil and culture in Venezuela, focusing in particular on the enduring persistence of the 'to sow the oil' slogan as a driver of policy making.

Chapter 2 provides a review of the relevant literature that forms the theoretical premises of this book, divided into four parts. Part one explores Henri Lefebvre's and David Harvey's theorisation on space, to develop a discussion on the production of space and State Space in order to focus on Lefebvre as a theorist of State Space as territory. Part two develops a discussion on state theory and Bureaucratic Power, to focus in particular on rentier state theory in order to define the particular characteristics of the Petrostate. Part three reviews key literature from the field of urban sociology to differentiate the terms city and urban, and their relationship with space and culture, to understand the effects of oil capitalism in the production of urban society and culture in the context of the Venezuelan Petrostate. Finally, it reviews relevant literature on the cultural dimension of oil, as well as the role of oil in and within culture, to examine the spatial and cultural representations of Petro-Socialism.

Chapter 3 examines the centrality of territory in the transition towards the Socialist State guided by the principle of the New Geometry of Power (the Fourth Engine of the Bolivarian revolution) with the creation of new policy instruments to reconfigure the national territory as a Socialist State Space. It traces the process of abrogation and substitution of the legal framework of political-administrative territorial management set up in the 1980s. It describes how Chávez's discourse informed the creation and implementation of new spatial strategies outlined in policy instruments created between 2005 and 2010 as a means of devising new spatial policies to dismantle the existing institutional apparatus of urban governance. This process was fraught with inconsistencies that opened a legal vacuum that diminished State Space authority and enabled PDVSA La Estancia to establish that the Oil Social Districts defined by the new

Organic Law of Hydrocarbons superseded the authority of regional and municipal governments.

Building on the previous discussions, Chapter 4 deploys a critical discourse analysis framework focused on the relationship between power, discourse and performative utterances to examine public speeches of the three leading figures of the national oil industry between 2005 and 2014: President Hugo Chávez, former president of PDVSA Rafael Ramírez and former General Manager of PDVSA La Estancia, Beatrice Sansó de Ramírez. It demonstrates that Chávez did not envision a post-oil world; on the contrary, his model relied on the expectations of an inexhaustible supply of oil rent that assured the endurance of Petro-Socialism and the Socialist State. Thus, Ramírez's speeches discursively establish PDVSA's identity as a revolutionary oil corporation; he instrumentalises his share of Bureaucratic Power as the head of the state-owned oil company to contribute to Chávez's vision of Petro-Socialism. In turn, Sansó de Ramírez fleshes out two discursive strands: the first, PDVSA La Estancia is an instrument of the Sowing Oil Plan that 'harvests culture' and second, the 'utopia of the possible'. This chapter draws on Zygmunt Bauman (2004) and Jeremy Ahearne (2009) to demonstrate how these discursive constructions, built on the stratum of the disjointed process to constitute the Socialist State Space, ultimately enabled PDVSA La Estancia to interpret Article 5 of the Organic Law of Hydrocarbons as an instrument of implicit cultural policy.

Chapter 5 explores the discursive construction of Culture as Renewable Oil of the advertisement campaign launched by PDVSA La Estancia in 2013 titled 'We transform oil into a renewable resource for you', featuring giant oil workers, through the semiotic lens of Charles S. Peirce's semiosis and Roland Barthes' Mythologies. The giant oil worker functions as an indexical sign of PDVSA, their inclusion and interaction with the spaces depicted in the adverts visually reframes them as oil fields in a clear attempt at naturalising a direct and mechanistic relationship between oil, urban space and culture, functioning also as a visual metaphor of PDVSA's State Space. The giant oil worker metaphorically transforms oil into culture. The analysis draws on George Yúdice's expediency of Culture as a Resource to argue that PDVSA La Estancia discursively renders oil and culture equivalent by evoking a farming cycle ('PDVSA La Estancia is oil that harvests culture') that encapsulates the discursive strands of 'renewable oil' oil and 'utopia of the possible' to depict a novel dramaturgical act of the New Magical State: Culture as Renewable Oil, as such it is tied to the land; thus territory and culture become indivisible. Hence, Culture as Renewable Oil becomes inextricable from the Oil Social District as PDVSA's parallel State Space. If culture can be 'harvested' from the subsoil, then the Petrostate can claim complete ownership and tight control over culture as a 'renewable resource' as established by the Organic Law of Hydrocarbons.

Finally, the Conclusion returns to the discussions developed in the individual chapters and locates them within the historical imperative to 'sow the oil' and the unravelling of Petro-Socialism, modelled on unrealistic expectations of enduring high oil revenues. It summarises that in a Petrostate, oil binds Territory,

Bureaucratic Power and Culture, and makes wider points regarding the Venezuelan state-owned oil company's ownership and authority over city spaces, bolstered by its direct access to the oil revenue and the fragmentation of the bureaucratic structure of the state apparatus. In the particular case of PDVSA La Estancia, the notion of Culture as Renewable Oil, personified by the giant oil workers in the adverts, negate the original political, economic and cultural processes that brought to fruition the public art and architectural structures depicted, for they were produced by a state that was considered by Hugo Chávez as bourgeois, capitalist and counter revolutionary.

References

Ahearne, J. (2009) 'Cultural Policy Explicit and Implicit: A Distinction and Some Uses', *International Journal of Cultural Policy*, 15(2), pp. 141–153.

Atehortúa Cruz, A. L. and Rojas Rivera, D. M. (2005) 'Venezuela antes de Chávez: auge y derrumbe del sistema de "Punto Fijo"', *Anuario Colombiano de Historia Social y de la Cultura*, 32, pp. 255–274.

Baptista, A. (1997) *Teoría económica del capitalismo rentístico*, 2nd edn. Caracas: Banco Central de Venezuela.

Barrett, R. and Worden, D. (eds) (2014) *Oil Culture*. Minnesota: University of Minnesota Press.

Bauman, Z. (2004) 'Culture and management', *Parallax*, 10(2), pp. 63–72.

Beblawi, H. and Luciani, G. (1987) 'Introduction', in Beblawi, H. and Luciani, G. (eds) *The Rentier State*. London: Croom Helm, pp. 1–21.

Brenner, N. and Elden, S. (2009a) *Henri Lefebvre. State, Space, World: Selected Essays*. Minneapolis: University of Minnesota Press.

Brenner, N. and Elden, S. (2009b) 'Henri Lefebvre on State, Space, Territory', *International Political Sociology*, 3(4), pp. 353–377.

Coronil, F. (1997) *The Magical State: Nature, Money, and Modernity in Venezuela*. Chicago: University of Chicago Press.

Coronil, F. (2000) 'Magical Illusions or Revolutionary Magic? Chávez in Historical Context', *NACLA Report on the Americas*, XXXIII(6), pp. 34–42.

Coronil, F. (2011) 'Magical History What's Left of Chávez?', in *LLILAS Conference Proceedings, Teresa Lozano Long Institute of Latin American Studies*. Latin American Network Information Center, Etext Collection.

Corrales, J. and Penfold, M. (2011) *Dragon in the Tropics. Hugo Chávez and the Political Economy of Revolution in Venezuela*. Washington D.C.: The Brookings Institution.

Darwich, G. (2008) 'Institucionalidad petrolera en Venezuela de 1959 a 1963: entre continuidades y discontinuidades', *Cuadernos del CENDES*, 25(67), pp. 35–58.

González Casas, L. and Marín Castañeda, O. (2003) 'El transcurrir tras el cercado: ámbito residencial y vida cotidiana en los campamentos petroleros en Venezuela (1940–1975)', *Espacio Abierto*, 12(3), pp. 377–390.

Harvey, D. (2006) 'Space as a Keyword', in Castree, N. and Gregory, D. (eds) *David Harvey: A Critical Reader*. London: Blackwell Publishing Ltd, pp. 270–293.

Harwich Vallenilla, N. (1984) 'El Modelo Económico del Liberalismo Amarillo, historia de un fracaso, 1888–1908', *Universidad Santa María, Centro de Investigaciones Históricas*.

Jessop, B. (1990) *State Theory: Putting the Capitalist State in Its Place*. Cambridge, UK: Polity Press.

Joyce, P. and Bennett, T. (eds) (2010) *Material Powers: Cultural Studies, History and the Material Turn*. London: Routledge.

Karl, T. L. (1997) *The Paradox of Plenty: Oil Boom and Petro-States*. Berkeley: University of California Press.

Kozak Rovero, G. (2014) 'Cultura en la ley: nación, pueblo, historia y democracia en la Revolución Bolivariana', *Anuario Ininco/Investigaciones de la Comunicación*, 26(1), pp. 319–340.

Kozak Rovero, G. (2015) 'Revolución Bolivariana: políticas culturales en la Venezuela Socialista de Hugo Chávez (1999–2013)', *Cuadernos de Literatura*, XIX(37), pp. 38–56.

LeMenager, S. (2014) *Living Oil. Petroleum Culture in the American Century*. Oxford, UK: Oxford University Press.

López-Maya, M. (2003) 'The Venezuelan "Caracazo" of 1989: Popular Protest and Institutional Weakness', *Journal of Latin American Studies*, 35(1), pp. 117–137.

López-Maya, M. (2006) *Del viernes negro al referendo revocatorio*, 2nd edn. Caracas: Alfadil Ediciones.

López-Maya, M. (2007) *Nuevo debut del Estado mágico, Aporrea.org*. Caracas, Venezuela. Available at: www.aporrea.org/actualidad/a35326.html (accessed: 8 December 2015).

Lord, B. (2014) *Art & Energy: How Culture Changes*. Chicago: University of Chicago Press.

Maass, P. (2009) *Crude World*. London: Allen Lane.

McGuigan, J. (2003) 'Cultural Policy Studies', in *Critical Cultural Policy Studies: A Reader*. Malden, MA: Blackwell, pp. 23–42.

Miller, P. and Rose, N. (2008) *Governing the Present: Administering Economic, Social and Personal Life*. Cambridge: Polity.

Miller, T. and Yúdice, G. (2002) *Cultural Policy*. London: Sage Publications Ltd.

Mitchell, T. (2011) *Carbon Democracy: Political Power in the Age of Oil*. London: Verso.

Mommer, B. (1994) *The Political Role of National Oil Companies in Exporting Countries: The Venezuelan Case*. Oxford: Oxford Institute for Energy Studies.

Mukerji, C. (2010) 'The Unintended State', in *Material Powers: Cultural Studies, History and the Material Turn*. London: Routledge, pp. 81–101.

O'Brien, D. (2014) *Cultural Policy: Management, Value and Modernity in the Creative Industries*. New York: Routledge.

Pérez Schael, M. S. (1993) *Petróleo, cultura y poder en Venezuela*. Caracas: El Nacional.

Quintero, R. (2011) 'La cultura del petróleo: ensayo sobre estilos de vida de grupos sociales de Venezuela. [The Culture of Oil: Essay on the Life Styles of Social Groups in Venezuela]', *Revista BCV*, pp. 15–81.

Salamanca, L. (1994) 'Venezuela. La crisis del rentismo', *Nueva Sociedad*, 131, pp. 10–19.

Silva-Ferrer, M. (2014) *El cuerpo dócil de la cultura: poder, cultura y comunicación en la Venezuela de Chávez*. Mexico City: IberoAmericana.

Straka, T. (2016) 'Petróleo y nación: el nacionalismo petrolero y la formación del estado moderno en Venezuela (1936–1976)', in Straka, T. (ed.) *La Nación Petrolera: Venezuela, 1914–2014*. Caracas, Venezuela: Universidad Metropolitana, pp. 105–168.

Tinker Salas, M. (2014) *Una herencia que perdura, petróleo, cultura y sociedad en Venezuela*. Caracas, Venezuela: Editorial Galac.

Urbaneja, D. B. (2013) *La renta y el reclamo: ensayo sobre petróleo y economía política en Venezuela*. Caracas, Venezuela: Editorial Alfa.

Uslar Pietri, A. (1936) 'Sembrar el petróleo', *AHORA*, 14 July.

Uslar Pietri, A. (2001) 'De una a otra Venezuela', in Arráiz Lucca, R. and Mondolfi Gudat, E. (eds) *Textos Fundamentales de Venezuela*. Caracas: Fundación para la Cultura Urbana, pp. 285–306.

Uzcátegui, R. (2010) *Venezuela: La Revolución como espectáculo. Una crítica anarquista al gobierno bolivariano*. Caracas: El Libertario.

Yúdice, G. (2003) *The Expediency of Culture: Uses of Culture in the Global Era*. London: Duke University Press.

1 Entanglements of oil, modernity, state and culture in Venezuela

Up until the late 1980s, Venezuela was considered exceptional in the region, as it was regarded as the most stable democracy and the most developed and wealthiest country in Latin America, all thanks to its oil industry. To unpack the intersections between statecraft, modernity and culture in Venezuela it is crucial to understand them as inseparable from the oil-based Bureaucratic Power of the state.

Venezuela did not become a rentier state with the rise of the oil industry in the early twentieth century, it was born a rentier state from the moment it became a republic by adopting the mining codes inherited from the Spanish Crown. The way the country dealt with the oil industry was akin to how it dealt with modernity through the use of archaic bureaucratic instruments which, combined with the state's technological and financial incapacities, translated into statecraft built around the notion of property and control over oil rent money. Oil was understood predominantly as wealth that flowed from the soil directly to the state's coffers and not as a complex technological reality.

The consolidation of state formation, modernisation and oil production in Venezuela, as well the state's modernising efforts, concentrated on the main urban centres, predominantly the capital, which broadly remains the domain of the elite. The transformation of the built environment through strategies of modernisation did not cascade into all sectors of Venezuelan society as oil created an inflated economy that fostered an illusion of progress and modernity promoted from the state, manifested more clearly in the rapid urbanisation in the twentieth century. Oil, the state and modernisation are interdependent phenomena in Venezuela, but tend to be predominantly studied within the disciplines of economics and politics; when reference is made to culture or urban development, these are regarded as the result of direct investment of the oil wealth made available by the state to private enterprise and altruistic elites.

This chapter traces state formation in Venezuela back to the postcolonial emergence of Latin American nations; it explains how the process of independence and the construction of bureaucratic structures was fraught with contradictions, such as the enduring rigid social structures inherited from the colony that perpetuated the social and political power of elites, at odds with the project of national modernisation. This chapter also provides a historical review of the political and cultural effects of the arrival of the oil industry in the early twentieth century up to Hugo Chávez's Petro-Socialism. The chapter develops in two parts. Part one contextualises state formation in Venezuela within the postcolonial process of the emergence of Latin American nation states in the nineteenth and twentieth centuries, addressing the enduring contradictions generated by a process of modernisation led by elites that came accompanied by the perpetuation of a colonial structure to preserve their economic and political power. It also explores the formation of the Venezuelan Petrostate from the period of post-independence to the arrival of the oil industry, and the emergence of what Venezuelan anthropologist Fernando Coronil has defined as the Magical State. Part two develops a discussion about the historical context of the intersections between statecraft, modernity and oil in Venezuela, exploring in particular the slogan 'to sow the oil' coined by Arturo Uslar Pietri in 1936, which runs across historical narratives around oil wealth and modernity, heightened by Carlos Andrés Pérez's The Great Venezuela project in the 1970s, the subsequent exhaustion of the democratic Pact of Puntofijo accelerated by the collapse in oil prices in the 1980s and 1990s, which laid the foundations for Hugo Chávez's Bolivarian revolution and the re-edition of the 'to sow the oil' imperative to build a new type of oil-based socialism.

Oil and the illusion of modernity

Latin American Spanish colonies achieved independence much earlier than many colonies in the global peripheries. According to Heinz Sonntag (1990, p. 405), these newly emerging republics adopted the institutions and constitutions of the most advanced models of the capitalist state in Europe. Intellectuals and elites involved in shaping the new nations imported European ideas and transplanted them in order to shape new state structures modelled after the European experience. Sonntag affirms that the path towards the consolidation of such young states was difficult because the recently independent nations lacked the internal socio-economic foundations that characterised the development of capitalist states in Europe. Latin American nations inherited rigid class structures and institutionalised relations of domination from the colonial period, as well as heavily dependent economies (Silva Michelena, 1971, p. 389) as the colonies were not seen as potential markets but merely as sources of wealth extraction.

The colonial policies of Spain were designed to consolidate Spain's monopoly over trade across the continent (Lynch, 1989, p. 142), secure dependence on the metropolis, and keep a tight grip on political rule by preventing its colonies from developing any national identity or independent trade. Spanish rule in the Americas was based on a balance of interest groups: the Church, the administration and local elites (p. 329). Spain almost exclusively appointed Spaniards to the higher ranks in political positions, leaving the minor political offices of the city open to *criollos* (Spanish Americans) which for centuries effectively prevented *criollos* from influencing local politics or developing associations of their own (Bethell, 1985, pp. 229–230; Lynch, 1985, pp. 9–10;

Carrera Damas and Lombardi, 2003). Colonial production was exclusively export oriented; raw materials such as silver, gold, coal and a variety of agricultural goods were shipped directly to the metropolis. Spain developed trading companies ventures, such as the *Real Compañía Guipuzcoana de Caracas* granted in 1728 which was given a monopoly of trade with Venezuela, turning the poor province into 'an export economy producing a surplus for the metropolis' and 'extending the monopoly to new privileged groups' (Lynch, 1989, p. 148; Silva Michelena, 1971, p. 389). However, these continental policies did not effect the colonies evenly, or prevent the colonies from developing economically and politically. On the contrary, major differences emerged depending on the 'natural and labour resources' available in each colony (Silva Michelena, 1971, p. 390). The Spanish colonies of Nueva España (Mexico), Perú, Alto Perú and parts of Nueva Granada (modern Colombia) that had gold and silver mines and a large indigenous labour force available developed first, becoming the political centres of the overseas territories (Adelman, 2006, pp. 61–64). The rest of the colonies devoted to the exploitation of 'exotic' tropical products, developed an economy of *haciendas* (plantations). The predominance of the hacienda economy in most colonies has been interpreted as the implementation of a feudal mode of production and social organisation, which is a misconception, considering the situation of Spain in the sixteenth and seventeenth centuries. It is more accurate to describe the modes of production in Spanish America as heterogeneous, where a diverse set of relations of feudal and mercantile production coexisted, with one key feature in common: the overexploitation of the labour force (pp. 58–61). The emphasis on satisfying the needs of the metropolis over the internal needs of the colonies fostered the formation of a particular racial class structure concentrated in regional enclaves that had more contact with the metropolis than with each other.

The solidity of the social, political and economic structures of the colonies began to weaken in the nineteenth century due to a concurrence of internal and external factors. Internally, increasing tensions between *criollo* landowners and Spanish merchants, the ambitions of *mestizos* and frequent revolts by Indians and slaves marked the progressive dissolution of the colonial structure (Adelman, 2006, pp. 218–219). Externally, *criollos* were not exempt from the global economic dynamics of the Iberian Atlantic trade in which they were immersed economically (Guerra, 1994, pp. 195–197; Adelman, 2006, pp. 101–140). British and Dutch contraband and piracy undermined Spanish control over its colonies, which was crippled by Napoleon's occupation of Spain. This environment made *criollos* all the more receptive to the ideas that had emerged from American Independence and the French Revolution, which helped launch the wars of independence.

Although the wars resulted in the political independence of the Latin American nations from Spain (Brazil's independence from Portugal and its postcolonial statehood followed a very different path), these wars were not catalysts for the development of an internal centralised bureaucracy. While war has been considered the key mechanism of bureaucratic centralisation in Europe, it was not the only way because 'war can even hinder bureaucratization'; there are a myriad of factors that are more influential than war such as 'elite ideologies, administrative models, religious doctrines, and elite politics' meaning that for postcolonial developing countries there is a long-standing debate between 'those who attribute the presence or absence of centralized bureaucracies to colonialism and those who offer alternative explanations' (Vu, 2010, pp. 151–152). The separation from the well-ordered and fixed Bourbon state, was followed by a turbulent transition towards a *caudillo* State 'where authority was personal and obedience unpredictable' (Lynch, 1992, p. 139). Although the wars of independence in Latin America led to the collapse of the colonial state, they were too short and isolated to have any cumulative impact on bureaucratisation (Lynch, 1992, p. 35; Vu, 2010, pp. 153–154), creating a vacuum that would be filled by competing social groups: politicians and *caudillos* (regional charismatic strongmen). At different moments, politicians and *caudillos* were part allies, part rivals; the *caudillo* is a 'child of war and a product of independence', and as elites in the capital were debilitated, the loss of control over the peripheries and provinces opened the door to an era of *caudillos* leadership and power (Lynch, 1992, pp. 34–36). Latin America's vast territory made bureaucratisation beyond the main cities extremely costly for their poor and dependent economies.

For the emerging nations and their governing urban elites, state building could not derive solely from territorial control and bureaucracy considering that ideas of administration, rights and rituals of rule played a much more important role (Vu, 2010, p. 165). In the case of Argentina and Venezuela the *caudillo* 'was usually the centre of a vast kinship group based on land and linking legislators, bureaucrats, and military in a powerful political network' who 'had some idea of the nation and its assets, which they monopolized and used as patronage to bind their small oligarchy ever closer together' (Lynch, 1992, p. 136). The *caudillos* 'became the state, and the state the property of the caudillo'; they were both agents and enemies of the nation state (Lynch, 1992, pp. 134–135):

> The caudillos jealously guarded their national resources, land and offices, for these were their stock of patronage, the assets on which ultimately their power was based. Caudillos could attract a necessary clientage by promising their followers appointments and other rewards when they reached power. And clients would attach themselves to a promising patron in the expectation of preferment when he reached the top. The system perpetuated personalism and retarded the process of state-building. For it was regarded as much safer to accept a personal promise from a caudillo than an anonymous undertaking from an institution, whether executive or legislative. The caudillo was real, the state a shadow. So the mutual needs of patron and client, founded on deeply rooted traditions of personal loyalty, were among the mainstays of caudillism in the new states as well as a source of social cohesion. They were also obstacles to political growth and allegiance to national institutions.
>
> (Lynch, 1992, p. 155)

Thus, the political history and bureaucratisation of independent Latin American nations has been shaped by 'political personalism understood as the personal exercise of power, as an expression of the pure will to dominate, subject to whim alone, concurrent with a general institutional instability and/or insufficiently rooted norms' (Soriano de García-Pelayo, 1993, p. 9, translation by the author). In Venezuela, the system of *caudillismo* was a response to a precarious state as it was considered a 'cheaper form of government' in which 'personalist solutions and informal expedients' took the place of formal instruments of power (Lynch, 1992, p. 210). The *caudillo* functioned as an ally of the debilitated urban elites in the development of the emergent national state. José Antonio Páez (1790–1873), a regional *caudillo* who fought with Simón Bolívar, led Venezuela's independence from Gran Colombia to become the first constitutional president of Venezuela in 1830. Páez 'led a coalition of Caracas landowners, merchants, and officials, which he held together on a platform of peace and security'; he identified with the 'the political, economic, and social interests of the Caracas agrarian and commercial groups' in opposition to agrarian and commercial groups of the llanos (Lynch, 1992, p. 175) who could potentially challenge the political hegemony of the elites in the capital.

The process of nation-building and state formation across Latin America developed unevenly, according to the peculiarities of each former colony. Those countries that inherited a consolidated political and economic colonial structure found them difficult to overcome; the colonies that had the most prosperous societies and the most integrated state–church apparatus 'were not the first to consolidate post-colonial state power' (Topik, 2002, p. 112). The development of an effective bureaucratic apparatus would be crucial to the rest of the continent as 'modern centralized bureaucracy is perhaps the most important institution in the structure of any state' (Vu, 2010, pp. 151–152). But it can hardly be said that Latin American states had achieved complete control over their territories; the history of the late nineteenth up until the mid-twentieth century accounts for the prevalence of *caudillismo*, a system in which *caudillos* used their own armed militias to overthrow government and take over power, severely undermining the constitution of state power and becoming one of the strongest obstacles to the development of the nation state (Lynch, 1992, p. 134).

The postcolonial period was predominantly defined by the pursuit of modernity with the adoption of modern political and economic techniques, but the modernity experienced in Latin America was peculiar; it combined novel discourses with inherited colonial structures (Lombardi, 1982, pp. 160–162; Zahler, 2013, pp. 49–53). Liberalism in Latin America played a crucial role in nation-building, elites attributed its set of ideas with the 'capacity to generate change, to bring about positive improvement in a society that was trying hard to "catch up" with the times' (Roldán Vera and Caruso, 2007, pp. 9–10). The need to 'catch up' with the times is a theme that would run across national state projects in the twentieth century, particularly in Venezuela, a predominantly agrarian country that upon the discovery of oil fields and the development of the oil industry at the turn of the twentieth century, had access to an unprecedented source of wealth which determined its swift transformation into an oil rentier economy and an ever more powerful state; the notion of 'backwardness' is key to understand how the oil rent was instrumentalised to 'catch up' with modern times (Baptista, 1999). But this also created an ambivalent relationship with oil (Bhabha, 1994, pp. 95–96), regarded both as a carrier of modernity and predator of national sovereignty and cultures.

Modern institutions struggled to take root in Latin America because they were incompatible with the colonial legacy, which generated a pessimistic view among elites on whether modernity could find fertile ground (Ortiz, 2002, pp. 252–253). Renato Ortiz (2002) and Tomás Straka (2006) address this incompatibility by speaking of a 'modern tradition' in Latin America. For Ortiz, tradition 'is everything that is inserted in daily culture' (2002, p. 258), a concept that is accepting of modernity which is no longer seen as a radical exclusion of the past, just as tradition and past are no longer identified with the exclusion of the new. Similarly, Venezuelan historian Tomás Straka speaks of a *modernidad criolla* (creole modernity) that emerged around the eighteenth century along with the consolidation of the *criollo* elite, among which the ideas of the Enlightenment circulated. *Criollos* thought of themselves as Europeans, they sought to be acknowledged as 'second Europeans' building a Young Europe in America (Straka, 2006, p. 23) rather than 'second-class Europeans'. *Modernidad criolla* was Eurocentric; its discourse shaped by the European project of modernity. The national project put forward by the *criollos* also implied the incorporation of their non-western subordinates – blacks, Indians and *mestizos*, regarded as barbaric enemies of civilisation – and the rest of the territory to their own modern logic and order (p. 39). Changes came in modern form – as in 'icons of progress' – but would transform neither institutions nor society so as not to challenge the colonial foundations of the political and economic power of *criollo* elites (p. 17). Modernity was conceived as the final stage in the transit towards complete westernisation. According to Straka, the national modern project of the *criollos* was defined by the following three characteristics:

> first, it is a continuity of the conqueror in its imposition of Western order on the New World; second, through the incorporation of subordinates into their logic, subordinates are creolised which in the course of the following two centuries would produce enough hybridisations for something completely new to emerge (…), and third, it is intertwined with all European and that, since the eighteenth century, is the modern corollary, in consequence its tradition, always to assume the novelties coming from Europe, it is enforced: it becomes the tradition of modernity.
>
> (Straka, 2006, p. 8, translation by the author)

The contradictions between *modernidad criolla* and the processes of modernisation are noted by Néstor García Canclini (1989, p. 42), he argues that while the 'second Europeans' of the nineteenth century and the new economic and political elites of the twentieth century wished to modernise their nations based on the European modern project, the discrepancies between modernity and modernisation were instrumental in preserving their status as dominant classes. The oligarchies of the turn of the century did not constitute States, they brought order to some areas of society, which created uneven and fragmentary development (García Canclini, 1989). By the late twentieth century Latin America had become a continent where 'traditions have not yet disappeared and modernity has not completely arrived' (p. 1). In the particular case of Venezuela, it wasn't until the rise of the oil industry in the early twentieth century that Venezuela acquired the economic and political resources to consolidate as a modern state with a centralised bureaucracy.

The manner in which Venezuela dealt with the rise of the oil industry was tantamount to how it dealt with modernity (Pérez Schael, 1993, p. 39). Venezuela's elites confronted modernity and the oil industry based on the traditions inherited from Spanish colonial systems and post-independence *caudillismo*; despite becoming an independent republic in 1830 the consolidation of a nation state was far from a lived reality (Baptista, 1997, p. 132).

At the dawn of the twentieth century Venezuela was a 'heavily indebted country, in deep political turmoil and plagued by military unrest' (Mommer, 1994, p. 27). The country remained a predominantly rural society, its economy and exports relied predominantly on the traditional monoculture of coffee, vulnerable to the fluctuations of the international market; by 1920 only 17 per cent of the population lived in urban settlements with only 2 per cent of the population involved in salaried labour in the manufacture of basic food products (Baptista, 1997, p. 131). The emerging Venezuelan state declared itself the heir of the rights of the Spanish Crown: ecclesiastic patronage, property over vacant lands and ownership over all mines (Pérez Schael, 1993, p. 39); the property of the Sovereign was based on the principle that what belonged to no one must belong to the King. This process was not exclusive of Venezuela, it was characteristic of all Latin American emerging states who transitioned towards a modern apparatus of governance.

After independence, the new republic substituted the King and inherited his rights, according to the terms sanctioned by Simón Bolívar in his 1829 Quito Decree, which validated the *Ordenanzas de la Nueva España* (Ordinances of New Spain) of 1783, until new mining laws were created. The colonial legislation was progressively dismantled throughout the twentieth century; the French model of legislation, adopted during the regime of Guzmán Blanco, also governed matters related to the oil industry in later decades. However, subsequent mining codes and the 15 new constitutions created up until 1936 maintained the same principles established by Simón Bolívar (Pérez Schael, 1993, p. 40). The wealth extracted directly from the subsoil in the form of rent became an affirmation of sovereignty as the rent sanctioned the recognition of the authority of the nation as analogous to the King's. The difference was that the nation did not have the personal incarnation of majesty of a king, it acted through the state as a mediator between national sovereignty and citizens (p. 40). The principle of sovereignty manifested exclusively through the power to grant property rights to the land and administer the rent; in the case of Venezuela it was the state as landlord that mediated this relationship. Thus, Venezuela did not become a rentier state with the rise of the oil industry, it was born a rentier landlord state from the moment it became a modern republic: 'rentierism does not derive from petroleum, on the contrary, the traditional legislation of mining was the instrument used to codify the specificity and complexity of oil, until it was converted into a prisoner of the rent' (p. 41). As the sovereignty of the state was built on the notion of land property, the rent annulled the mineral materiality of oil.

Venezuela lacked the resources and the capacity to exploit and produce oil products, the only option for the state was to continue the system of concessions of strategic investment to foreign companies established in the nineteenth century (infrastructure, rail, shipping, etc.); revenue from oil would come primarily in the form of concessions and royalties (Harwich Vallenilla, 1984). By losing its mineral materiality, oil would be reduced to rent money that flows from the soil directly to the state's coffers; oil mattered as money, not as a complex technological new reality (Pérez Schael, 1993, p. 94). The intellectual class attributed an ephemeral quality to oil because, unlike agriculture, the wealth produced relied on money and not on its mineral materiality; discussions around oil centred on issues of sovereignty and how to position oneself in regards to the territory oil wealth was extracted from: nationalism or treason (p. 95). Hence, for the Venezuelan state, the only matter to resolve was where, how and to who distribute oil money, not to how to produce it. Venezuela's identity as a nation became closely entangled with oil; as the Petrostate engaged with modernity and its power increased, Venezuelan society learned to see itself as an oil nation and to view the state as the single representative of a population unified by oil (Coronil, 1997, p. 84).

The relationship between oil, culture and modernity in Venezuela is underpinned by an enduring narrative of 'to sow the oil', a slogan originally coined in 1936 by Venezuelan intellectual, writer and politician Arturo Uslar Pietri. The phrase 'to sow the oil' headlined Pietri's seminal essay published in 1936 (Pietri, 1936) in which he defended the idea of taking advantage of the knowledge, technology and financial power of foreign oil corporations. Pietri saw great benefits in keeping the country open to foreign capital and to use this capital to invest in development and industrialisation[1] (Uzcátegui, 2010, pp. 37–38). He colluded farming and mining language as a didactic device to explain how oil wealth should be invested, making direct reference to the land from where riches were extracted and not to oil as an immaterial and ephemeral source of wealth.

For landowners and traditionalist elites, the only way to produce enduring wealth was agriculture. Oil was money that spurted from the entrails of the land, it could only provide ephemeral wealth and destruction, akin to a 'black gold rush' (Straka, 2016, pp. 131, 137). Landowners' apprehensions towards and rejection of oil were grounded on the invisibility of crude oil; it is a material entity hidden in the subsoil, its potential yield wasn't as visibly quantifiable as land and hectares of crops on the surface. Even more, the volume of oil reserves was not known with precision, underscoring the fears that oil deposits would soon be exhausted and bring ruin. It wasn't until 2011 (during the Chávez regime) that OPEC would certify that Venezuela possessed the largest oil reserves in the world. But most importantly, oil extraction challenged the grip of landowners on political-economic power over land, ideologically tied on their defence of a national identity deeply rooted in agriculture (the predominant mode of production since colonial rule). This was not surprising considering that Venezuela was still a rural nation dependent on an agricultural economy. Between 1920 and 1950, approximately 60 per cent of the Venezuelan population lived in the countryside (Straka, 2016, p. 134). This is the backdrop of Uslar Pietri's imperative 'to sow the oil', to use the wealth produced by a destructive and ephemeral mineral economy to develop and modernise the agricultural economy (Uslar Pietri, 1936; Straka, 2016, p. 139), under the leadership of an illustrated elite. By 1949 Uslar Pietri had identified an emerging 'feigned nation' with a fake modern scenography built upon transient oil, that once exhausted would lay bare the poor 'real nation' made of tinsel and still backward (Uslar Pietri, 2001; Straka, 2016, p. 140). To 'sow the oil' is at the centre of enduring conflicting views around oil; Uslar Pietri would revisit and replay this phrase in public debate throughout the twentieth century, in 1936, 1945, 1961, 1980, 1983 and 1990, claiming every time that oil had yet to be sown (Pérez Schael, 1993, pp. 199–205), proving that 'to sow the oil' remained an achievable but elusive utopia. It would be inaccurate to say that Venezuela never 'sowed' oil, the Petrostate did invest in modernisation, infrastructure and industrialisation, but its elusiveness was rooted in a deficient state apparatus and a political class unwilling to carry out necessary structural reforms.

When foreign oil companies began to establish in Venezuela in the early twentieth century, the material effects of oil wealth were not immediately felt on the areas surrounding the oil drills and refineries; they materialised first in the remote oil camps and the fenced residential quarters built by foreign oil corporations, as enclaves of modernity frequently surrounded by poverty belts (González Casas and Marín Castañeda, 2003, p. 381). The iron fences built to isolate the oil camps did not stop poor neighbouring communities peeping into the modernity of foreign capital: technology, urban planning, architecture, corporate culture and lifestyle (González Casas and Marín Castañeda, 2003, pp. 381–382). Rodolfo Quintero, a Venezuelan Marxist anthropologist and former oil camp dweller wrote in 1968 (2011) *La cultura del petróleo: ensayo sobre estilos de vida de grupos sociales en Venezuela* ('The Culture of Oil: Essay on the Life Styles of Social Groups in Venezuela', recently re-edited by the Venezuelan Central Bank). Quintero defined the 'culture of oil' as a foreign culture, a force of conquest with its own technology, instruments, inventions, equipment and non-material devices such as language, art and science that deteriorate local and indigenous cultures (2011, pp. 19–20). According to Quintero, the way of life of the culture of oil is characterised by the exploitation of the national oil wealth by way of monopolistic foreign companies. Venezuela, an underdeveloped country, was at a disadvantage in the face of the powerful colonising forces of North American oil corporations. Quintero viewed the arrival of the oil industry as a second wave of colonisation in which the oil tower substituted the wooden plough brought by Spanish conquerors (2011, p. 24). He unequivocally demonises the United States, oil wealth, rapid urbanisation and bureaucratic and technological apparatuses as the predators and destroyers of national culture, dividing the history of Venezuela into two main periods: the pre-oil era and a culture of oil era, materialised in two new urban developments, the oil camp and the oil city (2011, p. 25).

Quintero characterises the oil camp as a 'colonial institution' (Quintero, 2011, p. 26) governed by a foreign company located in a remote foreign metropolis; the oil camp is an instrument of foreign capitalists that creates and maintains a rigid class structure of exploiters and exploited, hierarchically sustained by managers and administrators. The oil camps built by foreign oil corporations, such as Creole, Shell and Mobil, transferred modern capitalist ways of building and living to Venezuela. The oil camps disturbed the life of the human groups of surrounding local communities of small towns and villages, profoundly transforming their economic, cultural and social landscapes. The oil camp's accumulation of capital, workforce and land overpowered the much weaker regional economies, many communities abandoned agriculture altogether to become a reserve of low skilled workers economically dependent on the oil camp (Quintero, 2011, p. 26).

The oil camp was a self-contained entity of the modernity of oil capitalism; oil cities developed around their fringes, some emerging anew while others grew out of small villages. In contrast to the oil camp, an oil city was improvised and monotonous, it lacked public services and institutions that fostered any active urban life (Quintero, 2011, pp. 46–48, 55). Neither national nor regional governments invested in their infrastructure even though the 1961 census revealed that approximately a quarter of the national population lived in oil cities (Quintero, 2011, p. 51); improvements such as paved streets, churches and schools were built by the foreign oil companies. For this reason, Quintero argued that an oil city could not produce art, science or any form of intellectual culture as all that prevailed was the business and 'culture of oil' (2011, p. 55). This notion is relevant for this book, because Quintero regards the 'culture of oil' and the spaces it produces as sterile, incapable of art, science or any form of intellectual or cultural production.

With the oil camps, a new social organisation emerged comprising capitalist bureaucrats, foreigners and nationals, who became the representatives of the 'culture of oil'. The Venezuelan born and raised oil bureaucrats are labelled by Quintero as 'Shell Men' or 'Creole Men'; euphemisms for those who live by and for the oil companies, who think and live like foreigners with the customs and consumption habits of the United States (Quintero, 2011, p. 40). The lifestyles of the 'Shell Men' and the 'Creole Men' were expressions of progress and modernity; the 'Shell Men' and the 'Creole Men' viewed themselves as culturally superior to the locals' 'primitive' ways (Quintero, 2011, p. 40). But Venezuela was not particularly more backward or advanced than other Latin American countries, and what Quintero highlights is an enduring narrative (Tinker Salas,

2014, p. 64) that has shaped the discourse about Venezuela's backwardness in the 'pre-oil era' in comparison to the sleek modernity displayed by the oil camps and oil cities.

The participation of the new Latin American nation states in global capitalist trade deepened existing asymmetries between capital cities (urban commerce) and the interior (rural production) (Rangel, 1984, p. 329). While urban societies modernised by adopting European and North American lifestyles and technologies, in rural regions many of the habits and behaviours established by Spanish rulers in the sixteenth century still endured; this would become a recurrent theme for Latin American literature and intellectual production, for whom the control capital cities exercised over trade of raw materials and agricultural goods rendered them outward looking 'macrocefalic' parasites of the native hinterland (p. 329). Hence, affirms Carlos Rangel, the Cuban Revolution was the revenge of rural Cuba over La Habana's 'parasitic' modernity (p. 330). In this same vein, Quintero displays a similar abject rejection of modernity brought about by oil camps; beyond his Marxist stance, his rejection is characteristic of colonial ambivalent subjects (Bhabha, 1994). On the one hand, oil is a curse for Venezuelans while on the other, it could potentially make every Venezuelan very rich (Torres, 2012). Quintero is part of a generation of intellectuals and writers (Mariano Picón Salas, Briceño Iragorry, Alberto Adriani, Arturo Uslar Pietri, Ramón Díaz Sánchez, Enrique Bernardo Núñez, Juan Pablo Pérez Alfonzo) who regarded oil as a colonising and destructive force, a calamitous matter that once extracted from the subsoil decimated an idealised agrarian lifestyle; authenticity was attached to the land, what was built over it (cities, oil fields, oil camps…) was secondary, derivative, unreal (Torres, 2009, pp. 89–93).

Quintero's ideas condense the mindset shared by many intellectuals throughout the twentieth century for whom oil, beyond a source of energy and wealth, had become a demonic force (Pérez Schael, 1993, p. 9). In Venezuelan literary fiction, in the novel *Mene* by Ramón Díaz Sánchez, the 'end of oil' is regarded as a happy ending for it allowed the 'blonde man' – Shell and Creole men – to leave so waters could become pure again (Pérez Schael, 1993, p. 147). The modernity brought by oil was viewed through a moral light that construed oil as the destructive agent of nature and culture. And here lies a primordial ambivalence behind the national paradigm: a lingering nostalgia about a lost agrarian past that created a discursive tension of simultaneously embracing and demonising the modernising magic powers of oil, as it created prosperity and poverty in the same degree.

While the state used the oil riches to remodel itself as a modern institution, the modernisation of the urban landscape was, to great extent, the consequence of the settlement of oil corporations. The influence of oil wealth in the consolidation of the state and urban modernisation in Venezuela is not just an outcome of capital investment. Changes in the *Ley de Hidrocarburos* (Law of Hydrocarbons) made by President Medina Angarita in 1943 consolidated the state as the owner of the subsoil as well as protector of the nation's oil and established that 10 per cent of all oil extracted had to be refined locally.[2]

Venezuelan statecraft became inextricable from the development of the oil industry. The establishment of international oil corporations enabled the state to expand its jurisdiction beyond the private sector by creating a permanent dominance over the public sector, comparable in the region only to Cuba (Karl, 1997, p. 90), meaning that a landlord state that owned the oil in Venezuela owned the country without challenge; for the first three decades of the twentieth century Juan Vicente Gómez embodied the landlord state. The first oil wells were drilled in the early 1910s during the dictatorship of Juan Vicente Gómez, whose regime granted concessions to foreign corporations to explore, extract, refine and commercialise Venezuelan oil (Sullivan, 1992, p. 259). Within a decade, Venezuela's economy had transformed from agricultural production by private enterprise to oil exploitation by foreign oil corporations, with the state acting as landlord. Oil became the vehicle for modernisation and industrialisation (Bye, 1979, p. 59; Coronil, 1997, p. 4). By 1929 the country was the largest oil producer in the world (Bye, 1979, p. 59).

The sudden wealth produced by oil exploitation in the early twentieth century internally divided Venezuela into two bodies, a political body and a natural body (Coronil, 1997; Almandoz Marte, 2000). These two bodies were 'a political body made up of its citizens and a natural body made up of its rich subsoil' (Coronil, 1997, p. 4). The state found in the concentration of oil wealth the instrument to affirm its authority, and though growing separate from the nation, the state materialised as the sole mediator between the political body of the nation and the natural body of oil, granting itself the power to transform the country into a modern nation by manufacturing ambitious projects of infrastructure (Coronil, 1997, p. 5). As the Petrostate consolidated its authority and became entangled with oil, the monopoly over the nation's oil wealth transformed the state into the 'single agent endowed with the magical power to remake the nation' (Coronil, 1997, p. 4). Fernando Coronil's *The Magical State: Nature, Money and Modernity in Venezuela* (1997) argues that the Venezuelan state exercised this monopoly dramaturgically, enacting 'collective fantasies of progress' by way of spectacular projects of development and infrastructure to seize its subjects through the power of marvel rather than with the power of reason: 'the state seizes its subjects by inducing a condition or state being receptive to its illusions – a magical state' (1997, p. 5).

Coronil acknowledged that the Magical State (Coronil, 2011) was inspired by the work of playwright and critic José Ignacio Cabrujas. In an interview for the magazine *Estado y Reforma* (1987), a tri-monthly publication of the Presidential Commission for the Reform of the State (COPRE), Cabrujas defined the Venezuelan state as a State of disguise (*Estado de disimulo*). It is the *caudillo* who decides what the state is and what the law should be, transforming the state into a 'legal trick' that justifies whims, arbitrariness and other forms of 'I do as I please' (*me da la gana*). This is the way the Venezuelan state behaves in a nation in a permanent state of becoming with precarious institutions. The expansion of the oil industry promoted the concentration of power in the presidency as the embodiment of the 'magical' powers of oil; the Magical State is personified by the president as a magnanimous sorcerer. Prior to the industrialised exploitation of oil, the state in Venezuela had remained an unfinished project that lacked a 'national army or an effective bureaucracy' with 'partial dominion over the nation's territory and sway over its citizens' (Coronil, 1997, p. 76). As oil wealth increased, so did the state's capacity to construct itself as a national institution, able to expand the range of its dominion through material signs of progress. With massive works of infrastructure and the vertiginous modernisation of cities in just a few decades, oil wealth created the illusion that modernisation could be achieved almost overnight (p. 68). Coronil's Magical State expands the Marxist analysis of Latin America, which is often centred on the relationship between labour and capital, and tends to overlook the role of land as the incarnation of all the powers of nature (Coronil, 2011, p. 4). The state's dominion over land and subsoil is not limited to the availability of resources, 'land is the foundation of both the Venezuelan state and Venezuelan society' (p. 5); the exploitation of land and its subsoil means power over the extraction and circulation of wealth because power over the land equates to power over oil, as both a natural resource and a political tool.

While, post-independence Venezuela was born a rentier state, with the exploitation of oil the country inaugurated its modern history as a Petrostate, 'one whose capacity to create consensus and enforce collective decisions rested largely on the fate of the international oil market as well as on its ability to tax foreign firms and distribute its gains' (Karl, 1997, p. 91). The state's revenue was not the product of internal national productive sectors, payment of the rent to the state came from the great margins of profitability of the oil business enjoyed by foreign corporations; hence, it is more accurate to say that a rentier landlord state 'captures' the rent, rather than produces its own rent (Baptista, 1997, p. 67; Urbaneja, 2013, p. 117). The political economy of the country functioned in terms of claiming and distributing oil rent, gradually becoming a society of what Diego Bautista Urbaneja defines as 'rent claimants':

> sectors of Venezuelan society do not *seek* rent, in the sense of putting themselves in the position of obtaining rents, but they *claim* portions of a rent that is already there and to which they have – or believe to have – rights to.
> (Urbaneja, 2013, p. XXII, translation by the author, emphasis in original)

This set in motion a historical cycle of widespread 'political rent-seeking' and 'rent claimant' behaviour manifested in the centralisation of authority in the executive (particularly in the presidency) and the continuous search for increased oil revenues through a deepening of oil dependence combined with the emergence of new popular demands to the state to amend the resulting imbalances (Karl, 1997, p. 91; Urbaneja, 2013, pp. 79–89).

The Venezuelan Petrostate, continuing its rentier landlord state character, rather than symbolise national glory came to be viewed mainly as an 'enormous distributive apparatus' of oil rent, a paternal state that provides but whose power is hollowed out by a gap between authority and territory. Oil wealth suddenly made possible lavish and monumental works of infrastructure (Coronil, 1997, p. 76) for a country that had been in chronic debt and lacked basic infrastructure, such as a national road network and communications systems.

After the 1920s, Venezuela became extremely attractive for foreign capital because it had one of the most liberal oil policies in Latin America, bringing a significant increase in the investment on infrastructure, urban development and architecture (González Casas and Marín Castañeda, 2003, p. 379). Juan Vicente Gómez's death in 1935 put an end to 27 years of dictatorship and consolidated 'the final breakthrough of capitalism as the dominant mode of production' (Bye, 1979, p. 63). The dominance of foreign capital fostered the emergence and growth of a 'petty bourgeoisie' long before a working class became relevant in numbers or an organised workforce. In addition, oil production initiated a process of fast urbanisation accompanied by the deterioration of the agricultural sector that had previously dominated the economy and social order, encouraging mass migration from rural areas to cities and centres of oil production in search for employment (Bye, 1979, p. 63; Fossi, 2012, p. 106), transforming the pattern of population settlement across the country. Furthermore, the increase of the state's fiscal income from the oil revenue enabled, for the first time in the country's history, ambitious programs of development of infrastructure essential for the modernisation of society in education, sanitation, healthcare, transportation and communication, aligned with the continuation of the creation of new legal instruments, institutional apparatus and centralised bureaucracy that began in the 1930s, to administer and govern the nation and its increasingly more urbanised society (Brewer Carías, 1976, p. 214; Fossi, 2012, pp. 107, 110).

The period of democratic transition, initiated by the presidency of General Medina Angarita (1941–1945), cut short by a coup that established the Revolutionary Junta government led by Rómulo Betancourt in alliance with the military (1945–1948, the *Trienio Adeco*), put forward the institutionalisation of the public function of urban planning at a national scale with the creation of the National Institute of Sanitary Works (Instituto Nacional de Obras Sanitarias – INOS) in 1943, attached to the Ministry of Public Works (Ministerio de Obras Públicas (MOP), funded in 1874) and The National Commission of Urbanism (Fossi, 2012, pp. 110–111). The Constituent Assembly of 1946–1947 repealed the Constitution of 1936 and approved the new Constitution of the United States of Venezuela of 1947, the first truly democratic constitution (Rey, 1991, p. 534), which consecrated the universal right to vote, widened the scope of political and social rights and delineated the direct role of the state in the nation's development and the solution of its social and economic problems (Planchart Manrique 1997). It was under this constitution that, in 1948, Acción Democrática's candidate Rómulo Gallegos, renowned novelist and politician, became the first civilian president to be democratically elected by secret, universal and direct vote, only to be overthrown ten months later by a coup d'etat led by the Armed Forces, castrating the democratic experiment to establish a dictatorial regime under a Military Junta Government, followed by Lieutenant Coronel Marcos Pérez Jiménez's rise as de facto president of Venezuela in 1952. Pérez Jiménez's regime established a developmental model guided by the philosophy of his New National Ideal: the radical transformation of the physical environment and the betterment of the moral, intellectual and material conditions of Venezuelans for an absolute possession of the territory. He invested the state's vast oil revenues in an ambitious programme of state modernisation and lavish urban infrastructure (Almandoz Marte, 2012, pp. 95–96). Pérez Jiménez's regime showed a particular interest in the modernisation of the administration of urban planning, adopting a functionalist and technocratic approach which oversaw the swift transformation of the urban landscape with modernist architecture and cutting-edge engineering works; Caracas in particular became the showpiece of the dictatorial regime's oil-based veneer of modernity and progress (Frechilla, 1994, p. 344; Plaza, 2008, pp. 1, 9–13; Almandoz Marte, 2012, p. 99).

In 1958, the overthrow of Marcos Pérez Jiménez's dictatorship marked the beginning of a new democratic era: *Puntofijismo*. The Pact of Puntofijo[3] or *Puntofijismo* refers to the governability alliance signed in the aftermath of the ousting of the dictatorship in 1958 by the three main political parties: Acción Democrática (AD) (statist, nationalist, populist), Comité de Organización Política Electoral Independiente (COPEI) (Christian democracy) and Unión Republicana Democrática (URD) (centre-left socio-economics, liberal politics); it excluded the Communist Party of Venezuela (Casanova, 2012, p. 4). *Puntofijismo* became a bi-partisan system where AD and COPEI alternated power for decades to come, after URD abandoned the alliance in 1960. The Pact of Puntofijo also operated as a lobby comprised by local business elite and corporate interests from the United States, with the aim of maintaining stability and creating consensus in the midst of entrenched disputes among political and social factors (Terán Mantovani, 2014, p. 128). To this end, the pact created what Juan Carlos Rey defines as a 'populist system of conciliation of elites', constituted by a complex system of negotiation and accommodation of heterogeneous interests through an array of informal rules and institutional arrangements to secure support and continuity for the emerging democratic regime (Rey, 1991, p. 543). The Pact of Puntofijo also established a new political order of national sovereignty based on the principle of *maximin*,[4] maximising consensus and minimising conflict while establishing democratic controls over the use of oil rent to increase revenue and redistribute the oil wealth among social sectors (Urbaneja, 2013, p. 195). The institutionalised access to oil wealth of the *Puntofijismo* Petrostate created mechanisms to placate demands from heterogeneous sectors and avoid conflicts by financing large public spending, without making compromises, affecting negatively certain interests or having to raise domestic taxes (Karl, 1997, p. 111); a new consensus ensued as broad popular sectors increasingly advocated for the nationalisation of the oil industry.

Rómulo Betancourt, of AD, was elected president in the general elections held in December 1958. A United Nations Technical Assistance Mission recommended to the new government urgent improvements to public administration, to modernise and professionalise bureaucracy 'in order to conduct economic and social development programmes' (Brewer Carías, 1976, pp. 214–215; Ochoa

Henríquez, 1996, p. 177). With this purpose the new government established the Public Administration Commission, 'presided over by scholars of repute in the field of administration', until 1974 (Brewer Carías, 1976, p. 215). The Commission put forward some changes to the upper levels of public administration and produced an Administrative Career Law project, submitted to Congress in 1959 but not approved (Brewer Carías, 1976, p. 215). The Commission also created the Central Office of Coordination and Planning of the Presidency of the Republic (CORDIPLAN), a central government agency of budgetary and administrative planning for economic and social development at a national scale, responsible for formulating the First Plan for the Nation, and subsequent national development plans until 1998 (Alegrett Ruiz, 1997). The First Plan focused on the industrialisation of the economy buttressed on the nation's mineral resources, with this aim, the Corporación Venezolana de Guayana (CVG) (Venezuelan Corporation of Guayana) and the Corporación Venezolana del Petróleo (CVP) (Venezuelan Corporation of Petroleum) were created in 1960. From 1962 onwards Betancourt's political support for administrative reform dwindled as 'efforts to maintain the democratic regime' took priority in the face of attempts from right-wing military groups and left-wing guerrillas to take power (Brewer Carías, 1976, p. 216).

Betancourt called back from exile Juan Pablo Pérez Alfonzo, one of the founders of AD and former Minister of Development under Betancourt and Gallegos, to lead the Ministry of Mines and Hydrocarbons. In the wake of Dwight Eisenhower's restrictions on oil imports put in place in 1959 (with negative impacts on the price of Venezuelan crude oil), Betancourt and Pérez Alfonzo deployed foreign policies that prioritised strategic alliances with oil producing countries in the Middle East and North Africa (Darwich, 2008, pp. 46, 53). Juan Pablo Pérez Alfonzo, a staunch nationalist, advocated for the state's absolute sovereignty over its oil reserves since he regarded oil as the 'devil's excrement' (2011). In 1960, Pérez Alfonzo and the Oil Minister from Saudi Arabia, Abdallah Tariki (Darwich, 2008, p. 51), along with Iraq, Kuwait and Iran, set up the Organisation of Petroleum Exporting Countries OPEC:

> For Venezuela, where a revolution had overthrown the military government and brought an elected government to power, the aim was to imitate the collective arrangement among US states for restricting production, in order to negotiate an increased share of oil revenues and conserve supplies, and thus to allow an orderly process of economic growth and avoid a premature depletion of reserves.
>
> (Mitchell, 2011, p. 167)

The belief that only the state could distribute the nation's oil wealth, and that such distributive capacity had to rise above partisan conflict was cemented in the new Constitution of 1961, which reaffirmed state intervention and turned the president into a 'supreme political arbiter' with very few mechanisms for accountability (Karl, 1997, p. 105). The new Constitution also changed the clauses of oil concessions; new ones needed the authorisation of both chambers of Congress while existing ones had to be handed over to the government. Foreign oil corporations decided to reduce investment, retreat from the oil camps, and repatriate many of their executives, an ominous event for the formerly prosperous oil cities whose inhabitants had accommodated their economic activities to the necessities of oil camp dwellers and employees.

Venezuela had given absolute freedom to foreign capital transactions, but this began to change in the 1970s. In 1973, the outbreak of another Arab-Israeli war and the oil embargo enacted by Arab oil producing countries generated a rapid fourfold increase in oil prices, followed by another rapid threefold increase in 1979 (Venn, 2013, para. 8.2); while this triggered the Energy Crisis for oil consumers, for oil exporting countries such as Venezuela it produced a significant increase in oil revenues. This windfall prompted the then leader of AD, Carlos Andrés Pérez, elected to the presidency in late 1973 with wide popular support, to promise that the increased financial power of the state would allow Venezuela to become a developed country in just a few years.

Fleeting mirages of oil wealth: from Pérez's Great Venezuela to Chávez's Petro-Socialism

The political rent seeking and rent claimant expectations explained earlier, defined the actions of Pérez's government. Pérez wanted to take part in building a new world order announced by the wave of nationalisations of key productive sectors in developing countries. His populist presidency was underpinned by his political charisma and personality cult, isolating himself from criticism by ruling by decree to concentrate Bureaucratic Power on the presidency; he used Bolivarian symbols to represent himself as a world leader of the Third World capable to seize the unique opportunity brought by the oil boom (Karl, 1997, pp. 122–130): 'Many citizens and much of the press viewed Pérez's success as the nation's success and his hopes as the nation's hopes' (Straka, 2017, p. 393). Pérez launched with little public debate the ambitious national development plan *La Gran Venezuela* (The Great Venezuela). At the centre of his megalomaniac plan were the nationalisation and expansion of the aluminium, steel and oil industries (Karl, 1997, p. 124). The state-owned oil company Petróleos de Venezuela Sociedad Anónima (PDVSA) was created by decree in 1975, taking over the functions and operations of the Corporación Venezolana del Petróleo (Darwich, 2008, p. 50). Pérez nationalised the oil industry in 1976, just one year after the promulgation of a new law that reserved to the state the industry and commerce of hydrocarbons, and the expropriation of iron and steel industries (Bye, 1979, p. 57). Pérez also put forward a new project of public administration reform to further modernise and professionalise an overgrown bureaucracy, building on Rafael Caldera's (1969–1974) Fourth National Plan[5] dovetailing of administrative reform and development, centred primarily on the efficiency and productivity of emerging state-owned companies like PDVSA (Brewer Carías, 1976, p. 221; Ochoa Henríquez, 1996, p. 179). In this context of wealth, Pérez decreed in 1976 the Law of Central Administration, created the Ministry of Environment and Renewable Natural Resources, the Ministry of Urban Development, and ratified the competencies of CORDIPLAN (Geigel Lope-Bello, 1994, p. 30). Furthermore, he also decreed the celebrations of the bicentenary of Liberator Simón Bolívar's birth, scheduled for 1983, with a lavish programme of celebrations and new works of infrastructure attuned to the country's oil wealth (Straka, 2017, p. 381). The unprecedented oil bonanza of 1973–1978 presented Pérez with a once in a lifetime opportunity to completely restructure the country, to 'catch up' at last, but it also entrenched deep social, economic and political contradictions that would prove everlasting.

In the years after the nationalisation of the oil industry, PDVSA projected an image of high efficiency that differentiated it from the rest of the public sector. Its managers, executives and workforce had been educated by the international oil corporations in their corporate culture and technocratic practices (Wainberg, 2004, p. 4; Tinker Salas, 2009). It was largely assumed that the oil policies established in the aftermath of the nationalisation were aligned with national interests. Political parties (including parties of the Left) and most importantly Congress 'lost interest in oil as a topic for debate and as a central feature of their programs' (Parker, 2006, p. 62). This allowed PDVSA to keep the oil company's policies on the margins of public debate. They pursued a policy of insulation from government interference on the grounds that it was serving public interest by protecting the state-owned oil company from endemic clientelistic practices in public administration to preserve it 'as an efficient modern corporation' (Parker, 2006, p. 63). This meant that the national executive, and particularly the Ministry of Energy and Mines, progressively relinquished their capacity to enforce policies ending up as 'mere rubber stamp[s]' of approval for the decisions of PDVSA (Parker, 2006, p. 62).

The signs of prosperity provided by the oil boom and Pérez's Great Venezuela cemented in the psyche of Venezuelan society the myth that the country was infinitely rich (Naím, 2001, pp. 20–21), and rapidly heading towards a stable and prosperous future, fulfilled by its vast oil reserves. A second oil boom in 1980 seemed to prove this to be true (Karl, 1997, p. 161). But by 1983, efforts in diminishing oil dependence on OPEC oil by industrialised countries in the aftermath of the Energy Crisis, the increase in oil production by non-OPEC countries and OPEC's inability to function as an effective cartel to stabilise oil prices pushed global oil prices down, plummeting from US$22.99 in 1983 to US$13.08 by 1986 (Karl, 1997, pp. 161–162), representing a dramatic plunge in oil revenue for Venezuela. Venezuela was considered to be one of the most stable democracies and one of the fastest growing economies in Latin America (Monaldi and Penfold, 2014, p. 285), but the decade that followed Pérez's Great Venezuela was one of gradual economic and political decline (López-Maya, 2006, p. 21; Urbaneja, 2013, p. 279). The surge in global oil prices in 1980 led the newly elected government of COPEI's Luis Herrera Campins (1979–1984) to increase public spending and foreign debt, and defer much needed reforms, a policy sustained even as oil revenues dipped, to cope with the ensuing fiscal crisis

(Karl, 1997, pp. 172–173). In 1983, the year of Simón Bolívar's bicentenary celebrations, Herrera Campins inaugurated the Metro subway system (began in 1976) and the Cultural Complex Theatre Teresa Carreño in Caracas, lingering symbols of The Great Venezuela's oil prosperity. But The Great Venezuela's promise of sustained oil-subsidised prosperity suffered its first blow on 18 February 1983, when Herrera Campins announced the devaluation of the national currency in the aftermath of a dramatic dip in oil prices in 1982, setting in motion a downward spiral of economic stagnation, high inflation and further increase of foreign debt, with the subsequent deterioration of quality of life for large sectors of the population (Salamanca, 1994, p. 11; López-Maya, 2006, pp. 22–23). This day would come to be known as *Viernes Negro*, Black Friday (López-Maya, 2006, p. 23) or in more colloquial terms as the 'end of the party'.

Amid this crisis that disrupted the mechanisms of channelling a now dwindling oil revenue for public spending, Jaime Lusinchi of AD was elected president in 1984. He acknowledged the urgent need of an integral reform of the state, for which he set up in 1984 the Comisión Presidencial para la Reforma del Estado (COPRE) (Presidential Commission for the Reform of the State) with the aim of constructing a 'modern, democratic and efficient state' with a frugal and austere public national administration (Ochoa Henríquez, 1996, p. 180; Banko, 2016, p. 166). COPRE produced a decentralisation strategy that included the transfer of powers to state governorships and 'direct election by secret ballot of mayors and state governors' (Banko, 2016, p. 166). In 1989 the strategy materialised in a set of laws: the Law of Decentralisation, Delimitation and Transfer of Competencies of Public Power, the Law of Election and Removal of Governors and reform to the Law of Municipal Regime that introduced mayors and their direct democratic election (Alvarez Itriago, 2010, p. 666). Although it was initially met with resistance by political leaders, 'decentralisation succeeded because it was basically a response to the gradual collapse of weak and inefficient state institutions' (Banko, 2016, pp. 166–167); between 1989 and 1999 decentralisation helped mitigate some of the political tensions and social conflicts that emerged after the *Caracazo* (Alvarez Itriago, 2010, p. 666). Nonetheless, the Petrostate's centralist fiscal strategies persisted, which undermined the efficiency and participatory capacity of local and regional governments.

Carlos Andrés Pérez was elected for a second presidency in 1988 with the hope he could bring back the golden days of The Great Venezuela, but his presidency was mired by economic, social and political turmoil (Atehortúa Cruz and Rojas Rivera, 2005, p. 264). On 16 February 1989, with barely a month in office, Pérez announced an IMF-backed programme of macroeconomic adjustments which most notably included a 100 per cent increase in the price of gasoline enforced over the weekend of 25 and 26 February (López-Maya, 2003, p. 120). A revolt known as the *Caracazo*, a country-wide popular revolt without precedence in Venezuela's contemporary history, intensified the economic and social crisis. Another political crisis unfolded in February 1992, when a small group of the army, with the support of leftist civilian groups, staged a failed coup d'état led by Lieutenant Coronel Hugo Chávez Frías (Coronil, 2000, p. 37; López-Maya, 2003,

p. 129); a second failed coup d'état took place in November 1992 led by officers from the Navy and the Air Force. Although Pérez's presidency survived both coups it did not survive the deterioration of his public image and political leadership; he was impeached by Congress and sentenced to house arrest by the Supreme Court in 1993 (Salamanca, 1994, p. 12). In parallel, the cycle of boom and bust in oil prices paired with deep economic and political crises meant that by the early 1990s PDVSA lacked 'sufficient capital to develop their country's oil and gas reserve base and the associated infrastructure, much less fund increasingly urgent economic development needs' (Wainberg, 2004, p. 4). To this end, PDVSA sought to establish itself as an international oil corporation and lobbied for tax reductions and limited fiscal obligations towards the state resulting in a legislative reform in 1993 that reduced PDVSA's tax burden (Parker, 2006, pp. 63–64). During Rafael Caldera's second presidency (COPEI, 1994–1999) PDVSA consolidated its plan to reopen the industry to foreign capital and increase productive capacity through the investment programme *Apertura Petrolera* (Oil Opening), in contradiction to OPEC's policy of maintaining oil prices by limiting production. The unexpected collapse of the Venezuelan banking sector in 1994 required massive bailouts from the government that further deteriorated the socio-economic landscape (Urbaneja, 2013, p. 343). Caldera's government relied on oil revenue to counterbalance the high cost of the bailouts, rooted in the belief that the oil industry alone would be sufficient to reactivate the economy; by the late 1990s PDVSA was accused of contributing to the country's financial crisis by decreasing its fiscal contributions (Parker, 2006, p. 63). Nonetheless, by exerting control over the main source of the state's income, PDVSA operated with a larger budget than the government, working like a 'prosperous first world company in an impoverished third-world nation' (Maass, 2009, p. 202). The *Apertura Petrolera* was implemented without much public debate; local political figures associated with nationalist leftist parties and a few organisations and associations linked to the oil industry were its most staunch critics, they rejected the plan for offering favourable conditions to foreign investors at the expense of state revenues and most importantly, for breaking with OPEC (Parker, 2006, p. 63; Urbaneja, 2013, p. 353). Among the most salient critical voices was former coup plotter and future president Lieutenant Colonel Hugo Chávez Frías.

These events signalled the exhaustion of the marriage between *Puntofijismo*'s populist system of elites conciliation, oil rentierism and rent claimant behaviour, leading to a crisis of the paternal relationship between Venezuelan society and the Petrostate (Torres, 2009, p. 104). The Venezuelan Petrostate, once a powerful force of modernisation and progress, no longer had the sufficient oil revenue to satisfy the increasingly urgent social demands of its rent claimant population. A Petrostate that was navigating the tortuous path of consolidating a political consensus to modernise the country and develop an efficient bureaucratic apparatus modelled on oil overabundance fatally collided with the fall in oil prices, a rent-seeking political class and widespread corruption which aggravated the gradual collapse of institutional stability and social welfare. The desired triad of oil, state and modernity was showing signs of severe deterioration.

Hugo Chávez, imprisoned after the failed coup d'état he led against president Carlos Andrés Pérez, was granted a presidential pardon by Rafael Caldera in 1994 which allowed him to launch his political career to run for the 1998 presidential elections as an outsider with his own political platform, the *Movimiento Bolivariano Revolucionario 200* (MBR-200) (Coronil, 2000, p. 36). Founded in 1983, the MBR-200 was a fusion of the military with groups of the radical left that up to that moment had been working clandestinely (Silva-Ferrer, 2014, p. 77). The *Apertura Petrolera* became one of the key issues of the 1998 presidential race (Wainberg, 2004, p. 6). Chávez condemned PDVSA for continuing to function like a 'state within the state'[6] (Párraga, 2010, p. 29; Giussepe Ávalo, 2014, p. 26) that breached OPEC policies contributing to the downfall of oil prices; by 1998 Venezuelan crude oil had reached a mere US$7.35 per barrel (Wainberg, 2004, p. 6).

Chávez campaigned as an outsider, on an anti-establishment political platform outlined as an alternative to neoliberalism (Parker, 2006, p. 64; Urbaneja, 2013, p. 362; Silva-Ferrer, 2014), tapping into widespread public rejection of *Puntofijismo's* political system. His presidential campaign promised to put a halt to *Puntofijismo* through a peaceful revolution that would radically change the political landscape of Venezuela (Coronil, 2000, p. 34). He won the presidential election by a landslide, which marked a pivot in Venezuela's political history and in the relationship between PDVSA and the state.

Chávez implemented reforms in the areas of social policy, development models and most importantly, oil policies (Parker, 2006, p. 64). Venezuelan historian Margarita López-Maya characterises Chávez's presidencies as the 'new debut of the magical state' (López-Maya, 2007; Coronil, 2011). She identifies close similarities between Chávez's government and the first presidency of Carlos Andrés Pérez in the centralisation of power and the use of the oil rent to transform the state: 'the Magical state seems to be taking possession once again of the body of the nation' (2007). But Chávez went further; he condensed all the bureaucratic powers of the state to embody a New Magical State in his persona. Chávez's government is better understood by viewing it in relation to Juan Vicente Gómez, Marcos Pérez Jiménez and particularly Carlos Andrés Pérez (Coronil, 2011, p. 3). Chávez, like those presidents before him, promised to use the oil rent to bring wellbeing and prosperity to the Venezuelan people (Coronil, 2000, p. 40), but in the context of hegemonic neoliberal globalisation he proposed a different model for modernity and development, bidding to completely refashion the country into a Socialist State (Coronil, 2011, p. 9).

The arrival of Hugo Chávez to the presidency also inaugurated a 'new and contradictory phase' (Silva-Ferrer, 2014, p. 24) for culture in Venezuela, the expansion of the cultural domain by the Bolivarian revolution aimed at increasing the number of direct beneficiaries of the Petrostate (Silva-Ferrer, 2014, p. 27). The transformations produced by the Bolivarian revolution led to the polarisation of culture as the reflection of the political polarisation of Venezuelan society as the state's cultural institutions were put at the service of the ideology of the Bolivarian project; the cultural apparatus of the state was ideologically reframed into a populist and clientelist instrument which put the 'notion of culture as a public service' in crisis (Silva-Ferrer, 2014, p. 24). The rise of the Bolivarian revolution by popular election (rather than by force) brought to power emerging social and political groups, meaning that this 'new' political and cultural landscape was not an abrupt break with the past but the product of 'intense and complex struggles for the redefinition of the social sphere, and in consequence, of the cultural sphere' (Silva-Ferrer, 2014, p. 25). The participatory, protagonic, multiethnic and pluricultural democracy proposed by Hugo Chávez further entrenched historical fractures that were at the root of the demise of *Puntofijismo*'s representative democracy.

To this end, one of the key promises of the presidential campaign was to reform the constitution through the creation of a Constitutional Assembly, which was approved by popular vote in April 1999. A second election, in July 1999, selected the individual delegates who would draft the new constitution. On August 1999, the Constitutional Assembly voted to grant itself the powers to abolish government institutions. A constitutional referendum on December 1999 approved with a wide margin the Constitution of the Bolivarian Republic of Venezuela (CBRV) which came into effect in 2000 (Tinker Salas, 2015, p. 137); it is the first constitution approved by popular referendum in the history of Venezuela. It displaced the Constitution of 1961 and inaugurated the era baptised by Chávez as the Fifth Republic. Among the innovations of the CBRV are the change of the name of the country from the Republic of Venezuela to the Bolivarian Republic of Venezuela, the extension of the presidential term from five to six years allowing for consecutive re-elections, and the reform of the structure of the government that established a decentralised government and granted greater powers to the legislative branch (King, 2013, p. 379), such as the reform of the institutional structure of the capital city, Caracas.

With a new constitution in place, a 'mega-election' for every elected official in the country was held in July 2000, in which Chávez won the re-election for a second term by a large margin. His re-election was followed by reforms in social policy, cultural policy, development models and most importantly, oil policies that 'sought to complete the nationalization of the oil industry' by closing the legal loopholes created by the 1976 law to exert complete control over PDVSA (Parker, 2006, p. 64; Tinker Salas, 2015). The economic policies of the first two years of Chávez's presidency focused on increasing oil revenue, strengthening Venezuela's position inside OPEC, re-establishing state control over PDVSA and reinstating the policy-making role of the Ministry of Energy and Mines (Wainberg, 2004, p. 6; Parker, 2006, p. 64). In November 2001 a new Organic Law of Hydrocarbons was promulgated that reduced taxes, increased royalties and 'mandated state possession of a majority of stocks in all mixed companies engaged in primary activity in the oil industry' (Parker, 2006, p. 65), strongly rejected by PDVSA's top management.

Conflict with PDVSA ensued. This was a confrontation with PDVSA's small elite of oil workers, the corporate culture of the Shell Men and Puntofijismo's oil-based model of modernisation, believed to have robbed the Venezuelan people of their right to have their slice of the oil rent. According to Chavistas[7] Marxist view, heavily influenced by the work of Rodolfo Quintero, the dominance of foreign oil companies was the real reason for Venezuela's economic backwardness (Gallegos, 2017, p. 84). Between 1999 and 2000 Chávez had appointed successive presidents challenging PDVSA's meritocracy which did little to change the company's corporate behaviour (Wainberg, 2004, p. 6; Parker, 2006, p. 65). The ongoing conflict evolved into direct confrontation between the oil company and the President. PDVSA's executives organised as *Gente del Petróleo* (People of Oil) joined forces with the opposition to overthrow Chávez, to take part in the one day general strike organised by FEDECAMARAS (Venezuelan Federation of Chambers of Commerce) and CTV (Confederation of Workers of Venezuela) in 10 December 2001, which set the stage for the short lived coup d'état against Chávez in April 2002. The confrontation continued in December 2002 when a majority of PDVSA's employees joined the Paro Petrolero (oil strike) which put the oil company at a standstill for 65 days (Wainberg, 2004, p. 6; Parker, 2006, p. 65). Once the government regained control of the oil company, Chávez dismissed 18,000 managers and engineers (Wainberg, 2004, p. 7; Parker, 2006, p. 65; Maass, 2009, p. 202). Reforms to PDVSA were implemented swiftly and with ease. This also inaugurated a gradual process of de-professionalisation of the state-owned oil company, severe disinvestment and most importantly a reversal of the bureaucratic modernisation started in the 1990s.

The overhaul of PDVSA's staff by Chávez meant that he could count with more loyal civil servants, which included naming his close ally Rafael Ramírez as its president, and his spouse, Beatrice Sansó de Ramírez as the General Manager of Centro de Arte La Estancia (now PDVSA La Estancia). He also increased the revenues to the state from 40 per cent to two thirds, and most importantly he shifted the institutional channels of the flow of the rent from PDVSA to the state: instead of transferring oil money to the government to be redistributed to the ministries that oversaw social programs, PDVSA was put in charge of new government programs, effectively transforming the oil company into the 'engine of revolutionary change' (Maass, 2009, pp. 202, 215) and opening a direct lifeline between PDVSA and social spending. Chávez, aware of the diminished capacity of the public sector, believed that 'an oil company would succeed where government ministries might not' (Maass, 2009, p. 215). In this sense, PDVSA went from being an autonomous state-owned entity to effectively becoming a 'parallel state'. The events of 2002 marked a turning point in Chávez's politics, with the radicalisation of his Bolivarian revolution to embark the nation on a transition towards Socialism (Coronil, 2011, p. 13). Chávez further altered the institutional channels of the rent flow from PDVSA to the state: instead of transferring oil money to the government to be redistributed to the ministries that oversaw social programs, with PDVSA as the key player in advancing his nationalist, anti-capitalist and anti-imperialist model. This shift was followed by an oil windfall, a new oil boom that surpassed that of 1970s, between 2003 and 2008 the price of oil 'was more than double the average' of

Chávez's first five years in the presidency (Corrales and Penfold, 2011, pp. 55–57); as with the previous oil boom this had a clear and palpable effects on public spending, the political instrumentalisation of PDVSA and the foundations of Chávez's political project onwards. In 2005 he launched the *Plan Siembra Petrolera* (Sowing Oil Plan), a 25-year national plan and oil policy that formed the foundation for the advancement of Petro-Socialism to lay the groundwork for the transition towards the Socialist State. Chávez adopted Uslar Pietri's use of farming language to refer to oil. PDVSA's *Plan Siembra Petrolera* – Sowing Oil Plan – is a direct reference to Uslar Pietri's 'to sow the oil', but with Uslar Pietri's focus on foreign investment, there is a conceptual contradiction in using it to name Chávez's nationalist and anti-imperialist development plan (Uzcátegui, 2010, p. 38). The use of farming language to refer to the activities of the oil industry (such as the national Sowing Oil Plan) is underpinned by the enduring myth that oil, like a seed, can be 'sowed', discursively suggesting a natural renewable farming cycle of sowing and harvesting the subsoil for oil wealth.

The re-election of Hugo Chávez for a third term in 2006 revealed the degree of radicalisation of his policies with the creation of the single government party United Socialist Party of Venezuela (PSUV) and the launch of the National Project Simón Bolívar First Socialist Plan 2007–2013 (PPS) along with the Five Motors of the Revolution. Chávez declared his third presidential term as the beginning of a new era, the era of the expansion of the Bolivarian Revolution towards Socialism as the only alternative for transcending capitalism: 'The people voted for the way of socialism and it is socialism that the people want, what the fatherland needs' (Chávez, 2007b, p. 63). He announced that he was building a socialism of the twenty-first century (2007a) and emphasised the uniqueness of his socialist project in an *Aló Presidente* (Chávez, 2007a), broadcasted from the Orinoco Oil Belt, where he affirmed that he was building a socialist model different to the 'Scientific Socialism' that Karl Marx had originally envisioned, a Bolivarian socialism supported by the oil rent, a *socialismo petrolero*, in other words, Petro-Socialism. Bernard Mommer, Vice-Minister of Hydrocarbons and also former Director of PDVSA UK and Venezuelan representative at OPEC was present in the broadcast; he affirmed on national TV that oil was a blessing for socialism because it provided the resources for an easier and accelerated advancement of the project (Chávez, 2007a). Broadly, Petro-Socialism is focused on using oil revenues to fund the transition towards a socialist state and a new socialist society. Petro-Socialism is an anti-neoliberal project that reinforced oil extractive enclaves, criticised global capitalism while actively engaging in the global oil market. In these terms, Petro-Socialism is a peculiarly extreme form of oil rentierism. Underpinned by a steady rise in oil prices, this new era of Petro-Socialism promised historically neglected social sectors that they would finally enjoy enduring prosperity provided by oil.

The transition to the Socialist State was outlined in the First Socialist Plan for the Economic and Social Development of the Nation 2007–2013, which included a five step strategy named the Five Motors of the Bolivarian Revolution, to create the Venezuelan socialist model (Chávez, 2007b). Chávez reinstated the Petrostate at the centre of his political project, presenting oil rentierism as the essential condition to lead Venezuela towards 'development' (Terán Mantovani, 2015, p. 112) and laid out his ambitions to transform Venezuela into a 'world energy power'. While, according to Chávez's narrative, the Fourth Republic[8] trailed the path to 'sow the oil' but failed, the Bolivarian revolution would achieve what previous governments could not: the 'harvest of oil'. This is spoken from the standpoint of a new political class that became the sole owner of the 'seed' (oil) vindicating the oil wealth for collective benefit like governments in the past (Urbaneja, 2013, pp. 81–89); but this 'harvest' of oil is an act of appropriating already existing oil wealth, not the production of new sources of wealth.

In 2007 Hugo Chávez put forward a referendum to amend the CBRV, coinciding with the advancement of Petro-Socialism. The amendment was conceived as an instrument for the dismantlement of the 'constitutional and legal superstructure' that had sustained the capitalist mode of production, in order to embark on the construction of a socialist society for the twenty-first century. Although the reform lost the referendum vote, the legal foundations for the Socialist State had already been laid out by the National Assembly in clear breach of the CBRV, with the sanction in 2006 of the *Ley de Consejos Comunales* (Law of Communal Councils), reformed and elevated to the status of constitutional law in 2009 (Brewer Carías, 2011, p. 127); the law inaugurated a process in which communal power would render obsolete parishes, councils, municipalities and its respective authorities (Chávez, 2007b, p. 69). In December 2010, a month before the newly elected National Assembly took power with a larger representation of the opposition, a number of organic laws were swiftly sanctioned (Brewer Carías, 2011, p. 128) to establish the legal framework of the Socialist State.

Rafael Ramírez, president of PDVSA (until his demotion in 2014 by President Nicolás Maduro) simultaneously occupied the posts of president of the state-owned oil company, Minister of Energy and Petroleum, Vice-President of Territorial Development of the Republic and President of the United Socialist Party of Venezuela (PSUV) (Párraga, 2010, p. 122; Colgan, 2013, pp. 205–206). Ramírez was put in charge of PDVSA's new 'social sense' (Párraga, 2010, pp. 24–26), to expand the oil company's functions beyond its core commercial mission of generating maximum oil revenue to the state:

> Ramírez explicitly stated, and Chávez agreed, that PDVSA employees owe political allegiance to the Bolivarian Revolution, and that they should vote for Chávez or leave their jobs. PDVSA was delisted from the New York Stock Exchange, so its accounting practices no longer needed to comply with international standards for transparency. Most importantly, the company's revenue and assets became freely accessible to the government, which uses them for expenditure programs, including the misiones bolivarianas. In

2005, the transfers from PDVSA to the *misiones* were almost seven trillion bolivars ($3.2 trillion), more than twice the financial contributions from the central government itself.

(Colgan, 2013, p. 206)

In order to fulfil PDVSA's new socialist sense, seven subsidiaries were created by Hugo Chávez's mandate. The principal one, PDVSA Agrícola (PDVSA Agricultural) was created to literally 'sow the oil'; PDVAL (PDVSA Foods) distributes food staples but unlike the state-funded popular food market, Mercal, it was aimed directly at the middle classes; PDVSA Servicios developed in partnership with the Belarusian oil company absorbed the seismic analyses previously commissioned to private contractors; PDVSA Industrial manufactures oil and non-oil related equipment; PDVSA Naval was created to expand the company's fleet; PDVSA Gas Comunal to supply domestic gas; and PDVSA Urban Development took over some of the responsibilities of the Ministry of Infrastructure, Housing and Habitat (Párraga, 2010, pp. 26–29). The new relationship between PDVSA and the government was summarised in the slogan 'PDVSA now belongs to all' to reflect the direct channelling of the oil rent into social investment (Corrales and Penfold, 2011, p. 83). In practice, ministries that had traditionally been in charge of social spending were replaced by PDVSA (Corrales and Penfold, 2011, p. 84).

Chávez went for a fourth re-election in the presidential campaign of 2012. Despite strong speculations around his poor health (he had been diagnosed with cancer in 2011) he pushed an aggressive campaign to defeat by a slim margin the opposition leader Henrique Capriles. But Chávez was not able to attend his inauguration in January 2013, as he was still recovering from cancer surgery in Cuba; Chávez passed away in March 2013. Vice President Nicolás Maduro became interim president until the new presidential elections held in April 2013. The PSUV unanimously appointed Maduro as their presidential candidate, who won with an even narrower margin to opposition leader Henrique Capriles. Meanwhile, while oil prices peaked in 2011, they began to steadily decline, crashing in 2014.

The death of Hugo Chávez in March 2013 left the transition towards the Socialist State orphaned of its leader and mastermind. By then, Venezuela was even more dependent on oil revenue than before, the 'government take exceeded 90 per cent and was considered one of the highest in the world'; the crash in oil prices in 2014 laid bare the deficiencies of the state's apparatus and that the country's economic growth was 'mostly the result of higher oil revenues, thanks to a global oil boom, than from Chávez's economic policies' (Gallegos, 2017, p. 84). Beyond a dramaturgical exercise of Bureaucratic Power, the control over PDVSA enabled Chávez to summon all the Bureaucratic Powers of the State in his persona, but as will be made clear in Chapters 3 and 4, by delegating to PDVSA many functions of the government, he paved the way for the Venezuelan oil company to exercise power as a parallel state.

Conclusion

The above discussion lays out the backdrop of the topics this book investigates: the discursive and spatial dimension of the entanglement between oil, territory Bureaucratic Power and culture in the Venezuelan Petrostate. Venezuela's statecraft, modernity and culture are deeply entangled with oil. Venezuelan society learned to see itself as an oil nation unified by oil, with the Petrostate as their single representative; the Petrostate exercises its monopoly over oil wealth as a Magical State, a dramaturgical exercise of Bureaucratic Power enabled by oil revenue managed as it were manna that flows from the soil to the state's coffers.

Chávez's ambition to transform Venezuela into a Petro-Socialist world energy power were formulated under the mirage of an everlasting oil windfall. Although this ambition discursively placed him alongside the previous regimes of Marcos Pérez Jiménez and Carlos Andrés Pérez and their modernising projects; Chávez's 'harvest' would translate, in the end, into the bankrupting of Venezuela's modernisation project. Chávez's New Magical State stood on an ephemeral foundation to build a new kind of Socialist State. To this end, Chávez transformed PDVSA into the engine of revolutionary change, handing over many government functions and new social programs, in the process eroding the bureaucratic structure consolidated thus far while transforming PDVSA into a parallel state enabled by the institutional fractures of the transition towards the Socialist State. This sets the background of this book's investigation on the interfaces between oil, modernity, statecraft and culture, inextricable from the Bureaucratic Power of the New Magical State. Hugo Chávez's Petro-Socialist project was an extreme form of oil rentierism that further entrenched political rent seeking and widened the base of rent claimants during times of oil boom; a model historically destined to its own exhaustion, the bust in global oil prices led to the inevitable collapse of Petro-Socialism's feigned prosperity.

Notes

1 Arturo Uslar Pietri would come back to the imperative to 'sow the oil' during his tenure as Senator for Congress in the 1960s and 1970s, in the midst of the emerging democracy of the Pact of Puntofijo and the oil windfall of the 1970s, his essays and speeches to Congress address the urgency of a rational investment of the oil windfall.

2 This required oil corporations to build refineries and undertake a massive transfer of technology and bureaucratic workforce, which made it necessary to set up local corporate headquarters (Vicente, 2003, p. 394). Foreign oil companies such as the Creole Petroleum Corporation, Royal Dutch Shell, Mobil and Atlantic settled their headquarters in the country's capital, Caracas. Creole's buildings were located in the neighbourhood of La Candelaria, an eastern extension of the colonial quarter to the east of the city centre. Shell built headquarters in San Bernardino, a recently built modern neighbourhood adjacent to La Candelaria designed by the French urbanist Maurice Rotival. The settlement of foreign and local oil enterprises consolidated these neighbourhoods as an Oil District (now contained within Libertador municipality), from where modern practices in urban planning and architecture would spread to the rest of the city and the country (Vicente, 2003, pp. 397–398). By the 1990s, the Oil District had transformed into the nation's political and financial centre. Currently, Libertador

municipality concentrates many public buildings such as the headquarters of PDVSA, the Central Bank of Venezuela, the Federal Legislative Palace (National Assembly), the National Archives, many ministries, as well as notable modern buildings such as the Parque Central Complex (which up until 2003 were the tallest skyscrapers in Latin America), the corporate towers of Mercantil Bank and BBVA Provincial Bank, and the ill-fated Tower of David.

3 Puntofijo was the name of Rafael Caldera's residence in Caracas.
4 For more on *maximin* see Urbaneja's *La renta y el reclamo. Ensayo sobre petróleo y economia política en Venezuela* (2013), pp. 195–277.
5 During Rafael Caldera's presidency all proposed administrative reform projects were turned down by Congress, with the exception of the Administrative Career Law approved in 1970 (Brewer Carías, 1976, p. 235).
6 Statement borrowed from Gonzalo Barrios (1902–1933), founding member of Acción Democrática, who in 1983 in his position as senator for the Congress of the Republic of Venezuela, qualified PDVSA as a 'black box' and a 'state within the state'.
7 Chavista refers to supporters of Hugo Chávez.
8 Hugo Chávez's new constitution of 1999 inaugurated what he defined as the Fifth Republic, to differentiate it from the Fourth Republic, the democratic period of the Pact of Puntofijo (1958–1998).

References

Adelman, J. (2006) *Sovereignty and Revolution in the Iberian Atlantic*. Princeton, NJ: Princeton University Press.

Alegrett Ruiz, J. R. (1997) 'Sistema Nacional de Coordinación y Planificación', in Rodríguez Campos, M., Colmenares, Sara, Castro, Álvaro García, González, Javier, Méndez Salcedo, Ildefonso and Pellicer, Luis Felipe P. (eds) *Diccionario de Historia de Venezuela. Tomo 3*, 2nd edn. Caracas: Fundación Empresas Polar. Available at: http://bibliofep.fundacion empresaspolar.org/dhv/entradas/s/sistema-nacional-de-coordinacion-y-planificacion/.

Almandoz Marte, A. (2000) *Ensayos de cultura urbana*. Caracas, Venezuela: Fundarte, Alcaldía de Caracas.

Almandoz Marte, A. (2012) 'Modernidad urbanística y Nuevo Ideal Nacional', in *Caracas, de la metrópolis súbita a la meca roja*. Quito: OLACCHI, pp. 95–101.

Alvarez Itriago, R. (2010) 'Perspectivas de la descentralización y la participación ciudadana en el Gobierno de Hugo Chávez (1999–2009)', *Revista de Ciencias Sociales (RCS)*, XVI(4), pp. 665–676.

Atehortúa Cruz, A. L. and Rojas Rivera, D. M. (2005) 'Venezuela antes de Chávez: auge y derrumbe del sistema de "Punto Fijo"', *Anuario Colombiano de Historia Social y de la Cultura*, 32, pp. 255–274.

Banko, C. (2016) 'Redefining Regional Policies in Venezuela', in Scott, J. W. (ed.) *Decoding New Regionalism: Shifting Socio-political Contexts in Central Europe and Latin America*, 2nd edn. New York: Routledge, pp. 161–175.

Baptista, A. (1997) *Teoría económica del capitalismo rentístico*, 2nd edn. Caracas: Banco Central de Venezuela.

Baptista, A. (1999) *Una historia que no se hizo historia. El siglo XX venezolano*. Caracas: Comisión V Centenario de Venezuela.

Bethell, L. (1985) 'A Note on the Church and the Independence of Latin America', in *The Cambridge History of Latin America Volume 3: From Independence to c.1870*. Cambridge: Cambridge University Press, pp. 229–234.

Bhabha, H. (1994) *The Location of Culture*, 2nd edn. New York: Routledge.

Brewer Carías, A. (1976) 'The Administrative Reform Experience in Venezuela 1969–1975: Strategies, Tactics and Perspectives', in Leemans, A. F. (ed.) *The Management of Change in Government. Institute of Social Studies (Series on the Development of Societies), Volume 1*. Dordrecht: Springer, pp. 213–237.

Brewer Carías, A. (2011) 'Las Leyes Del Poder Popular Dictadas En Venezuela En Diciembre De 2010, Para Transformar El Estado Democrático Y Social De Derecho En Un Estado Comunal Socialista, Sin Reformar La Constitución', *Cuadernos Manuel Giménez Abad*, 1, pp. 127–131.

Bye, V. (1979) 'Nationalization of Oil in Venezuela: Re-Defined Dependence and Legitimization of Imperialism', *Journal of Peace Research*, 16(1), pp. 57–78.

Cabrujas, J. I. (1987) 'Heterodoxia y estado: 5 respuestas', *Estado & Reforma*. Edited by COPRE, p. 129.

Carrera Damas, G. and Lombardi, J. (eds) (2003) *Historia general de América Latina, Volume 4*. Organización de las Naciones Unidas para la Educación, la Ciencia y la Cultura UNESCO.

Casanova, R. (2012) *Partidos políticos venezolanos: ideas para su reinvención*. Caracas, Venezuela. Available at: http://library.fes.de/pdf-files/bueros/caracas/09213.pdf.

Chávez, H. (2007a) *Aló Presidente #288, www.alopresidente.gob.ve*. San Diego de Cabrutica, Venezuela: Sistema Bolivariano de Comunicación e Información SIBCI. Available at: www.alopresidente.gob.ve/materia_alo/25/1396/?desc=nro[1]._288_alo_presidente_-_28-jul-2007__estado_anzoategui___corregido_ljc_.pdf (accessed: 28 August 2015).

Chávez, H. (2007b) 'Juramentación del Presidente de la República Bolivariana de Venezuela, Hugo Chávez Frías (período 2007–2013)'. Caracas, Venezuela: Asamblea Nacional de la República Bolivariana de Venezuela.

Colgan, J. (2013) *Petro-Aggression: When Oil Causes War*. Cambridge: Cambridge University Press.

Coronil, F. (1997) *The Magical State: Nature, Money, and Modernity in Venezuela*. Chicago: University of Chicago Press.

Coronil, F. (2000) 'Magical Illusions or Revolutionary Magic? Chávez in Historical Context', *NACLA Report on the Americas*, XXXIII(6), pp. 34–42.

Coronil, F. (2011) 'Magical History. What's Left of Chávez?', in *LLILAS Conference Proceedings, Teresa Lozano Long Institute of Latin American Studies*. Latin American Network Information Center, Etext Collection.

Corrales, J. and Penfold, M. (2011) *Dragon in the Tropics. Hugo Chávez and the Political Economy of Revolution in Venezuela*. Washington D.C.: The Brookings Institution.

Darwich, G. (2008) 'Institucionalidad petrolera en Venezuela de 1959 a 1963: entre continuidades y discontinuidades', *Cuadernos del CENDES*, 25(67), pp. 35–58.

Fossi, V. (2012) 'Desarrollo urbano y vivienda: la desordenada evolución hacia un país de metrópolis', in Almandoz Marte, A. (ed.) *Caracas, de la metrópolis súbita a la meca roja*. Quito, Ecuador: OLACCHI.

Frechilla, J. J. M. (1994) *Planes, Planos y Proyectos para Venezuela:1908–1958. Apuntes para una historia de la construcción del país*. Caracas: Fondo Editorial Acta Científica Venezolana.

Gallegos, R. (2017) *Crude Nation*. Lincoln: Potomac Books, University of Nebraska Press.

García Canclini, N. (1989) *Hybrid Cultures: Strategies for Entering and Leaving Modernity*. Minneapolis: University of Minnesota Press.

Geigel Lope-Bello, N. (1994) *Planificación y Urbanismo*. Caracas, Venezuela: Equinoccio, Universidad Simón Bolívar.

Giussepe Ávalo, A. R. (2014) *Visión petrolera de Hugo Chávez Frías. Teoría Socialista sobre la Política Petrolera Venezolana.* Caracas, Venezuela: Editorial Metrópolis.

González Casas, L. and Marín Castañeda, O. (2003) 'El transcurrir tras el cercado: ámbito residencial y vida cotidiana en los campamentos petroleros en Venezuela (1940–1975)', *Espacio Abierto*, 12(3), pp. 377–390.

Guerra, F.-X. (1994) 'La desintegración de la monarquía hispánica: revolución e independencia', in Annino, A., Castro Leiva, L. and Guerra, F.-X. (eds) *De los Imperios a las Naciones: Iberoamérica.* Zaragoza, Spain: IberCaja, pp. 195–227.

Harwich Vallenilla, N. (1984) 'El Modelo Económico del Liberalismo Amarillo, historia de un fracaso, 1888–1908', *Universidad Santa María, Centro de Investigaciones Históricas.*

Karl, T. L. (1997) *The Paradox of Plenty: Oil Boom and Petro-States.* Berkeley: University of California Press.

King, P. (2013) 'Neo-Bolivarian Constitutional Design Comparing the 1999 Venezuelan, 2008 Ecuadorian and 2009 Bolivian Constitutions', in *Social and Political Foundations of Constitutions.* Cambridge, MA: Cambridge University Press, pp. 366–397.

Lombardi, J. (1982) *Venezuela. The Search for Order, the Dream of Progress.* Oxford, UK: Oxford University Press.

López-Maya, M. (2003) 'The Venezuelan "Caracazo" of 1989: Popular Protest and Institutional Weakness', *Journal of Latin American Studies*, 35(1), pp. 117–137.

López-Maya, M. (2006) *Del viernes negro al referendo revocatorio*, 2nd edn. Caracas, Venezuela: Alfadil Ediciones.

López-Maya, M. (2007) *Nuevo debut del Estado mágico, Aporrea.org.* Caracas, Venezuela. Available at: www.aporrea.org/actualidad/a35326.html (accessed: 8 December 2015).

Lynch, J. (1985) 'The Origins of Spanish America Independence', in Bethell, L. (ed.) *The Cambridge History of Latin America Volume 3: From Independence to c.1870.* Cambridge, UK: Cambridge University Press, pp. 1–50.

Lynch, J. (1989) *Bourbon Spain 1700–1808.* Oxford: Basil Blackwell.

Lynch, J. (1992) *Caudillos in Spanish America 1800–1850.* Oxford, UK: Oxford University Press.

Maass, P. (2009) *Crude World.* London: Allen Lane.

Mitchell, T. (2011) *Carbon Democracy: Political Power in the Age of Oil.* London: Verso.

Mommer, B. (1994) *The Political Role of National Oil Companies in Exporting Countries: The Venezuelan Case.* Oxford: Oxford Institute for Energy Studies.

Monaldi, F. and Penfold, M. (2014) 'Institutional Collapse: the Rise and Decline of Democratic Governance in Venezuela', in Hausmann, R. and Rodríguez, F. (eds) *Venezuela Before Chávez: Anatomy of an Economic Collapse.* Pennsylvania: Pennsylvania State University Press, pp. 285–320.

Naím, M. (2001) 'The Real Story Behind Venezuela's Woes', *Journal of Democracy*, 12(2), pp. 17–31.

Ochoa Henríquez, H. (1996) 'La reforma de la Administración Pública en Venezuela – Proyectos y Realidad', *Gestión y Análisis de Políticas Públicas*, 7–8, pp. 177–188.

Ortiz, R. (2002) 'From Incomplete Modernity to World Modernity', in Eisenstadt, S. N. (ed.) *Multiple Modernities.* New Brunswick, NJ: Transaction Publishers, pp. 249–260.

Parker, D. (2006) 'Chávez and the Search for an Alternative to Neoliberalism', in Ellner, S. and Tinker Salas, M. (eds) *Venezuela: Hugo Chávez and the Decline of an Exceptional Democracy.* Plymouth, UK: Rowman & Littlefield Publishers, pp. 60–74.

Párraga, M. (2010) *Oro Rojo*. Caracas, Venezuela: Ediciones Puntocero.

Pérez Alfonzo, J. P. (2011) *Hundiéndonos en el excremento del diablo*. Caracas, Venezuela: Banco Central de Venezuela.

Pérez Schael, M. S. (1993) *Petróleo, cultura y poder en Venezuela*. Caracas, Venezuela: El Nacional.

Planchart Manrique, G. *et al.* (1997) 'Constituciones de Venezuela', in Rodríguez Campos, M., Colmenares, Sara, Castro, Álvaro García, González, Javier, Méndez Salcedo, Ildefonso and Pellicer, Luis Felipe P. (eds) *Diccionario de Historia de Venezuela. Tomo 1*, 2nd edn. Caracas: Fundación Empresas Polar. Available at: http://bibliofep.fundacionempresaspolar.org/dhv/entradas/c/constituciones-de-venezuela/ (accessed: 16 August 2018).

Plaza, P. (2008) 'La construcción de una nación bajo el Nuevo Ideal Nacional. Obras públicas, ideología y representación durante la dictadura de Pérez Jiménez, 1952–1958', in *Semana Internacional de Investigación Facultad de Arquitectura y Urbanismo*. Caracas: Universidad Central de Venezuela, Section HP–12, p. 24.

Quintero, R. (2011) 'La cultura del petróleo: ensayo sobre estilos de vida de grupos sociales de Venezuela', *Revista BCV*, pp. 15–81.

Rangel, C. (1984) *Del buen salvaje al buen revolucionario*, 2nd edn. Caracas, Venezuela: Monte Avila Editores.

Rey, J. C. (1991) 'La democracia venezolana y la crisis del sistema populista de conciliación', *Revista de estudios políticos*, 74, pp. 533–578.

Roldán Vera, E. and Caruso, M. (2007) *Imported Modernity in Post-Colonial State Formation: The Appropriation of Political, Educational, and Cultural Models in Nineteenth-Century Latin America*. Frankfurt am Main: Peter Lang.

Salamanca, L. (1994) 'Venezuela. La crisis del rentismo', *Nueva Sociedad*, 131, pp. 10–19.

Silva-Ferrer, M. (2014) *El cuerpo dócil de la cultura: poder, cultura y comunicación en la Venezuela de Chávez*. Madrid/Frankfurt am Main: Biblioteca IberoAmericana/Vervuert.

Silva Michelena, J. A. (1971) 'State Formation and Nation Building in Latin America', *International Social Sciences Journal*, 23(3), pp. 384–398.

Sonntag, H. (1990) 'Venezuela: el desarrollo del estado capitalista', in *El Estado en América Latina: teoria y práctica*. Mexico: Siglo Veintinuno Editories.

Soriano de García-Pelayo, G. (1993) *El personalismo político hispano Americano del siglo XIX*. Caracas, Venezuela: Monte Avila Editores.

Straka, T. (2006) 'La tradición de ser modernos. Hipótesis sobre el pensamiento criollo', in *La tradición de lo moderno: Venezuela en diez enfoques*. Caracas, Venezuela: Fundación para la Cultura Urbana, pp. 3–41.

Straka, T. (2016) 'Petróleo y nación: el nacionalismo petrolero y la formación del estado moderno en Venezuela (1936–1976)', in Straka, T. (ed.) *La Nación Petrolera: Venezuela, 1914–2014*. Caracas, Venezuela: Universidad Metropolitana, pp. 105–168.

Straka, T. (2017) 'La esperanza del universo: El bolivarianismo durante la "Gran Venezuela" (1974–1983)', *Revista de Indias*, LXXVII(270), pp. 379–403.

Sullivan, W. M. (1992) *Política y economía en Venezuela 1810–1991*. Caracas, Venezuela: Fundación John Boulton.

Terán Mantovani, E. (2014) *El fantasma de la Gran Venezuela: un estudio del mito del desarrollo y los dilemas del petro-Estado en la Revolución Bolivariana*, CLACSO. Caracas, Venezuela: Fundación CELARG. Available at: www.clacso.org.ar/libreria-latinoAmericana/libro_detalle.php?id_libro=1092&pageNum_rs_libros=0&totalRows_rs_libros=1057 (accessed: 22 March 2016).

Terán Mantovani, E. (2015) 'El extractivismo en la Revolución Bolivariana: "potencia energética mundial" y resistencias eco-territoriales', *IberoAmericana*, XV(59), pp. 11–125.

Tinker Salas, M. (2009) *The Enduring Legacy: Oil, Culture, and Society in Venezuela*. Durham: Duke University Press.

Tinker Salas, M. (2014) *Una herencia que perdura, petróleo, cultura y sociedad en Venezuela*. Caracas, Venezuela: Editorial Galac.

Tinker Salas, M. (2015) *Venezuela: What Everyone Needs to Know*. Oxford, UK: Oxford University Press.

Topik, S. (2002) 'The Hollow State: The Effect of the World Market on State-Building in Brazil in the Nineteenth Century', in Dunkerley, J. (ed.) *Studies in the Formation of the Nation-State in Latin America*. London: Institute of Latin American Studies, pp. 112–132.

Torres, A. T. (2009) *La Herencia de la Tribu*. Caracas: Editorial Alfa.

Torres, A. T. (2012) *El imaginario petrolero venezolano, Cineforo sobre Reventón II (Carlos Oteyza)*. Caracas, Venezuela. Available at: www.anateresatorres.com/?p=1175 (accessed: 19 July 2018).

Urbaneja, D. B. (2013) *La renta y el reclamo: ensayo sobre petróleo y economía política en Venezuela*. Caracas, Venezuela: Editorial Alfa.

Uslar Pietri, A. (1936) 'Sembrar el petróleo', *AHORA*, 14 July.

Uslar Pietri, A. (2001) 'De una a otra Venezuela', in Arráiz Lucca, R. and Mondolfi Gudat, E. (eds) *Textos Fundamentales de Venezuela*. Caracas, Venezuela: Fundación para la Cultura Urbana, pp. 285–306.

Uzcátegui, R. (2010) *Venezuela: La Revolución como espectáculo. Una crítica anarquista al gobierno bolivariano*. Caracas, Venezuela: El Libertario.

Venn, F. (2013) *The Oil Crisis*, 2nd edn. New York: Routledge.

Vicente, H. (2003) 'La arquitectura urbana de las corporaciones petroleras: conformación de Distritos Petroleros en Caracas durante las décadas de 1940 y 1950', *Espacio Abierto*, 12(3), pp. 391–414.

Vu, T. (2010) 'Studying the State through State Formation', *World Politics*, 62(1), pp. 148–175.

Wainberg, M. (2004) *From 'Apertura Petrolera' To 'Apertura Gas Natural'? The Case of Venezuela*. Austin: Centre for Energy Economics, The University of Texas, pp. 1–9.

Zahler, R. (2013) *Ambitious Rebels. Remaking Honor, Law and Liberalism in Venezuela, 1780–1850*. Tucson, A.Z: University of Arizona Press.

2 Oil in the intersection between Territory, Bureaucratic Power and Culture as a Resource

The emergence of modern statecraft in Venezuela is inextricable from the arrival of the oil industry; Venezuela is a country where oil has a historical clout as the carrier of modernity and political power. Oil, understood primarily as wealth that flows like manna from heaven to the state's coffers, made the Venezuelan state more powerful as it exercised its monopoly over the oil rent. Territory, Bureaucratic Power and culture are intrinsically interwoven by oil in a Petrostate, there is a spatial dimension to this entanglement that is the focus of this chapter, which establishes the three interconnected theoretical premises that underpin the book: State Space, Bureaucratic Power and Culture as a Resource.

The chapter is divided into four parts. The first part provides a discussion on Lefebvre's triad of space and David Harvey's matrix of spatiotemporality followed by a review of Neil Brenner and Steve Elden's reading of Lefebvre as a theorist of State Space as territory. It then develops a review of the literature on state theory, focusing primarily on Bureaucratic Power and rentier state theory to define the particular characteristics of the Petrostate. Part three subsequently builds on the discussion on State Space and oil rentier states to examine the intersections between the literatures on city, culture and oil. Finally, part four reviews the literature on culture as well recent texts from the emerging field of Energy Humanities that address the cultural dimension of oil, which provide the theoretical foundations to characterise the notion of Culture as Renewable Oil construed within Petro-Socialism.

State Space as territory

With the emergence of modern science, mathematicians and philosophers monopolised the conceptualisation of space as an abstract 'mental thing' divorced from reality and social life (Lefebvre, 1991, pp. 1–7). This mental space created an abyss between the 'space of the philosophers and the space of people who deal with material things' (Lefebvre, 1991, pp. 4–6). Mental space became the site of theoretical practice and the reference point of knowledge, space viewed as a neutral container of social relations completely detached from social practices. Alternatively, Lefebvre proposed that space has an active role in knowledge and in action, defining space as a concrete abstraction that involves mental abstraction and physical materiality; space becomes a concrete reality *through* and *in* social practices.

Lefebvre's thinking on space must be put into the historical context of 1960s–1970s France and the efforts of the French State to reform the practice of urban planning to develop alternatives to post-war functionalism. French planning institutions engaged in a process of institutionalisation of critique. The introduction of new procedures for the participation of inhabitants politicised its operations and stimulated the emergence of critical urban research, including Marxist research (Stanek, 2011, p. ix). Lefebvre's critique engaged not just with philosophy, but with sociology, architecture and urbanism; the development of his theory of the production of space is an extension of his philosophical thinking and his involvement in empirical studies for several French institutions as well as his close relationship with French architectural culture which included intense exchanges with planners, urbanists and architects (Stanek, 2011, pp. vii–ix).

Through these cross-disciplinary engagements, Lefebvre developed a qualitative approach focused on space as a lived experience, opposed to the abstract space of state planning and post-war functionalist urbanism; he shifted the focus from things *in* space to the actual processes of its production, the multiple social practices that produce it and the political character of the process of the production of space (Lefebvre, 1991, p. 37; Elden, 2004, p. 189; Stanek, 2011, p. ix). For Lefebvre, production carries a wider meaning than the mere economic production of things, the term involves the production of society, knowledge and institutions (Elden, 2004, p. 184).

Rather than disregarding notions of temporality and history by privileging space, Lefebvre advanced the idea that space and time appear and manifest as different but are indivisible (Elden, 2004, pp. 185–186). Every society (or Mode of Production in Lefebvre's terms) has historically produced its own particular space. Capital and space are social processes since space is the 'general form of social practice in capitalist modernities' (Stanek, 2011, p. xiii). Space is a social relationship, inherent to relationships of property and bound to the forces of production (Lefebvre, 1991, p. 85). The representations of the relations of production that contain within them power relations also occur in space, in the form of monuments, buildings and works of art (Lefebvre, 1991, pp. 31–33).

Lefebvre proposed a conceptual triad of interconnected realms for understanding space as a social product conformed by Spatial Practice, Representations of Space and Representational Spaces. Spatial Practice (perceived space) embodies the associations and interactions between daily life (human actions) and urban reality. It is revealed and deciphered through the routes, networks and flows that tie and connect the places of private life, work and leisure; this is an impersonal space comprised of the flows of energy, money, transportation, commodities, labour, etc. (Lefebvre, 1991, p. 38). A Spatial Practice entails cohesiveness but not necessarily a logical coherence. Representations of Space (conceived space) tend towards a system of intellectually worked verbal signs that belong to the domain of planners, urbanists, politicians, scientists and technocrats (Lefebvre, 1991, pp. 38–39); this is the Cartesian realm of maps, models, plans, blueprints and designs which are formed historically, informed by the knowledge and ideologies that exert a dominant force in the production of space in any society. Representational Spaces (lived space) or the experienced space is the space of human subjectivity superimposed to physical space through the use of symbols and images that have their point of origin in history. It is the dominated space that 'imagination seeks to change and appropriate' producing symbolic works (Lefebvre, 1991, p. 39, 41–42). The three categories are not hierarchically ordered as they remain in a state of continuous dialectical tension; they contribute to the production of space in different ways according to the society – Mode of Production – and historical period (Lefebvre, 1991, p. 46).

The transition from one model of society (Mode of Production) to another results in contradictions within the social relations which inevitably transforms and revolutionises space, resulting in the production of a new space (Lefebvre, 1991, p. 46). Space as a 'concrete abstraction' brings together physical, mental and social constructions which become material reality through human practice. David Harvey draws on Lefebvre to regard space as 'an active moment within the social process' (2006b, p. 77) given that capitalism has a very close relation to daily life, which is not separate from the circulation of capital. If capital produces space in its own image, and urbanisation is the physical framework for capital accumulation, then the study of the evolution of a particular city can provide a better understanding of the urban processes of capitalism (2006b, pp. 80–101).

Harvey identifies three ways in which space can be understood: Absolute Space, Relative Space and Relational Space (2006a, p. 272). Absolute Space is the space of Newton and Descartes, usually represented as a pre-existing grid independent of time and matter, it is fixed and measurable. As such, it is devoid of contradictions, uncertainties, ambiguities, open to human calculation. In geometric terms, it is the space of Euclid, the space of engineering practices and cartography. Socially, it is the space of private property and territorial boundaries such as administrative units, cities and states.

Relative Space is the space of Non-Euclidean geometries and Einstein. Relative space is twofold: it has multiple geometries whose measurements depend on the frame of reference of the observer as Relative Space is impossible to understand without time. Rather than speaking of space and time as separate, it requires speaking of space-time or spatiotemporality. In Relative Space, time is fixed while space is mutable according to certain observable rules. This is the space of the study of the flows of commodities, money, people, energy, etc. Each 'flow' or spatiotemporality demands a different framework of understanding, while their comparisons can reveal issues of political choice as well (Harvey, 2006a, p. 273).

Relational Space is closely associated with Leibniz. This notion of space proposes that space cannot be isolated from the processes that define it; the relational aspect means that 'processes do not occur *in* space but define their own spatial frame' (Harvey, 2006a, p. 273). Measurement and calculability become problematic in Relational Space but Harvey challenges the assumption that space-time can only exist if it can be quantified because there are processes within the social, cultural, political and mental dimension that, while elusive, can only be approached from a relational notion of space.

Harvey argues that space is not just absolute, relative or relational. He concurs with Lefebvre in that space can only be construed through human practices; hence it can become one, two or all categories at once depending on the circumstances. Harvey developed a three by three matrix (Harvey, 2006a, p. 282) that intersects his and Lefebvre's categories of space in Table 2.1.

Although Harvey acknowledges that the matrix has its limitations (2006a, pp. 281–284), the cross-relations between the categories and the diverse combinations that arise in their intersections enables the analysis of complex scenarios where the use of one simple set of categories would not be sufficient. Harvey's matrix of spatiotemporality underpins the approach of this book, as it enables us to break apart and characterise the entangled spatial dimensions of the topics that are the subject of analysis to reveal the spatial intersections between oil, city, policy instruments and the cultural work of PDVSA La Estancia in Caracas, as displayed in Table 2.2.

The analysis unfolds through Chapters 3, 4 and 5, following the categories of the matrix to guide the arguments. The spatiotemporal scope is circumscribed to the intersections between Harvey's Absolute and Relative Space and Lefebvre's triad of Material Space, Representations of Space and Spaces of Representation. The Venezuelan state's spatial machinations of Petro-Socialism are 'hidden' in the Absolute-Representations of Space of policy instruments, and most importantly, concealed in the Relative-Spaces of Representation of the adverts.

Caracas is the Absolute-Material Space that defines the geographical boundaries of the city. Absolute-Representation of Space is the space of the policy instruments of territorial and public administration, the legal entities of space, that are shaped by the discourses of Petro-Socialism, exerting a dominant force in the production of space in the transition towards the Socialist State. The policy instruments conceptualise the political and administrative boundaries of State Space authority as a manifestation of the Bureaucratic Power of Hugo Chávez and PDVSA. The public art and public spaces restored by PDVSA La Estancia are located in a Relative-Material Space as they serve as physical markers of the territorial appropriation of Caracas by PDVSA, which speaks of a dual occupation as the physical entities of the art works and public spaces end up located simultaneously in two superimposed State Spaces. Relative-Representations of Space is the space of the Organic Law of Hydrocarbons and the Oil Social Districts as oil-based conceptualisations of space, fixed in time but mutable in terms of the 'areas of influence' of the centres of oil extraction. The Absolute-Material Space of Caracas is conceptualised by the oil company as part of the Oil Social Districts by construing the headquarters of PDVSA as a centre of oil extraction and distribution of the rent, in which the city becomes an extension of the oil field. The Relative-Spaces of Representation locates the space represented in the adverts of PDVSA La Estancia's campaign 'We transform oil into a renewable resource for you'; the adverts function as containers of a 'lived

Table 2.1 David Harvey's matrix of spatiotemporality

	Material space (experienced space)	Representations of space (conceptualised space)	Spaces of representation (lived space)
Absolute space	Walls, bridges, doors, stairways, floors, ceilings, streets, buildings, cities, mountains, continents, bodies of water, territorial markers, physical boundaries and barriers, gated communities…	Cadastral and administrative maps; Euclidean geometry; landscape description; metaphors of confinement, open space, location, placement and positionality; (command and control relatively easy) – Newton and Descartes	Feelings of contentment around the hearth; sense of security or incarceration from enclosure; sense of power from ownership, command and domination over space; fear of others 'beyond the pale'
Relative space (time)	Circulation and flows of energy, water, air, commodities, peoples, information, money, capital; accelerations and diminutions in the friction of distance	Thematic and topological maps (e.g. London tube system); non-Euclidean geometries and topology; perspectival drawings; metaphors of situated knowledges, of motion, mobility, displacement, acceleration, time-space compression and distantiation; (command and control difficult requiring sophisticated techniques) – Einstein and Riemann	Anxiety at not getting to class on time; thrill of moving into the unknown; frustration in a traffic jam; tensions or exhilarations of time-space compression, of speed, of motion
Relational space (time)	Electromagnetic energy flows and fields; social relations; rental and economic potential surfaces; pollution concentrations; energy potentials; sounds, odours and sensations wafted on the breeze	Surrealism; existentialism; psycho-geographies; cyberspace; metaphors of internalisation of forces and powers (command and control extremely difficult – chaos theory, dialectics, internal relations, quantum mathematics) – Leibniz, Whitehead, Deleuze, Benjamin	Visions, fantasies, desires, frustrations, memories, dreams, phantasms, psychic states (e.g. agoraphobia, vertigo, claustrophobia)

Table 2.2 Matrix of spatiotemporality of the intersection between Territory, Bureaucratic Power and Culture

	Material space (experienced space)	Representations of space (conceptualised space)	Spaces of representation (lived space)
Absolute space	Caracas	Policy instruments of territorial and public administration – legal entity of space – informed by the discourses of Petro-Socialism and the transition towards the Socialist State (Bureaucratic Power)	
Relative space (time)	Public Art and Public Spaces intervened by PDVSA La Estancia, located in two superimposed State Space(s)	Law of Hydrocarbons and Oil Social Districts (The New Magical State)	The space represented in the adverts of the 'we transform oil…' campaign by PDVSA La Estancia (Culture as Renewable Oil)

space' that synthesises absolute and relative spaces. In the adverts, space is fixed in time but made mutable in perception and meaning framed by the state-owned oil company. The subjectivity of the General Manager of PDVSA La Estancia is superimposed to the Absolute-Material Space of Caracas, visually and verbally re-imagined and re-presented as an oil field that produces 'renewable oil' in the form of culture, what this book proposes as the notion of *Culture as Renewable Oil*.

The following paragraphs provide a review of Lefebvre's ideas on the relation between State and space, and Neil Brenner and Stuart Elden's reading of Lefebvre as a theorist of State Space as territory.

Just as Lefebvre construed space as a 'concrete abstraction', the state too is a concrete abstraction (Elden, 2004, p. 189). Lefebvre's notion of 'abstract space' defines a 'sociospatial organisation that is at once produced and regulated by the modern state' (Brenner and Elden, 2009b, p. 358). Lefebvre argues that abstract space *is* political, its production entails 'new ways of envisioning, conceiving, and representing the spaces within which everyday life, capital accumulation, and state action are to unfold' (Brenner and Elden, 2009b, p. 359) and is inherently geographically expansive. As the State 'engenders social relations in space' (Lefebvre, 2003, pp. 225–226), abstract space becomes political; political because it is both the product of conflicting social practices and the instrument of struggles and conflicts with the State imposing its own rationality to the chaotic relationships between social groups. Thus, political, economic and social hierarchies are represented spatially. The space created, meant to be political and regulatory, is both bureaucratising and bureaucratised (Lefebvre, 2003, pp. 243–245) to perpetuate and reproduce relations of domination in three dimensions: ideological (social), practical (instruments of action) and tactical-strategic (the subordination of a territory's resources).

Lefebvre identifies three moments in the relation between space and the state. First, the production of a physical space, the national territory, a space that can be 'mapped, modified, transformed by the networks, circuits and flows that are established with it' (Lefebvre, 2003, p. 224) such as transportation, infrastructure, commerce, etc. At the centre of the physical space of the state lies the city, the material space where human and political actions manifest. Second, the production of a social space, which constitutes the state itself, every institution possessing 'an "appropriate" space' constructed under a minimum of consensus around 'an (artificial) edifice of hierarchically ordered institutions, of laws and conventions upheld by "values" that are communicated through the national language' (Lefebvre, 2003, pp. 224–225). Third, being a social space, the state also composes a mental space that 'includes the representations of the State that people construct' which is not to be confused with social or physical space (Lefebvre, 2003, p. 225).

One of the foundations of Lefebvre's theorisation of the state and the relation between state and space is the State Mode of Production (SMP) which comprehends the historical central role of the state in the survival and perpetuation of capitalism (Brenner and Elden, 2009b, p. 359). The State Mode of Production develops spatial strategies to make capital accumulation and commoditisation possible. The State intervenes through 'diverse organisations and institutions devoted to the management and production of space' (Brenner and Elden, 2009a, p. 227). The state is both the agent and the guiding hand of the production of space; organised through rationality it appears homogenous and monotonous allowing the State to intervene and introduce its presence, control and surveillance throughout its physical space: 'Is not the secret of the State, hidden because it is so obvious, to be found in space?' (Brenner and Elden, 2009a, p. 228). Only the State has the capacity to manage space on a large scale. In countries where the State takes control of energy production, the State continued to install a dominant space 'extending the space demarcated by motorways, canals, and railroads' (Lefebvre, 2003, pp. 237–238).

Although Lefebvre did not explicitly theorise about territory, Neil Brenner and Stuart Elden (Brenner and Elden, 2009b) offer a reading of Lefebvre as a theorist of the state-space-territory dimension, or more specifically, of State Space as territory. Their point of departure is John Agnew's challenge to social scientists to overcome the three unquestioned geographical assumptions of the 'territorial trap' (Agnew, 1994, p. 59): first, states understood as entities of fixed sovereign space; second, domestic and international realms are neatly separated; and third, the state as 'container' of society. Simplistic conceptions of territory, such as defining the state as a fixed entity bound to a timeless territorial form put serious intellectual constraints in developing a spatial analysis of the state by assuming its territory as a 'self-evident' pre-existing dimension (Brenner and Elden, 2009b, p. 356; Agnew, 2010, p. 779). For Brenner and Elden, the insights of Lefebvre on territory overcome Agnew's 'territorial trap' by drawing on three conceptual points of State Space as territory: the production of territory, state territorial strategies and the territory effect (Brenner and Elden, 2009b, p. 353).

Brenner and Elden trace Lefebvre's notion of territory to a key passage in *The Production of Space* (1991) that refers to the semantic, historical and substantive relation between '*le terroir et le territoire*' (Brenner and Elden, 2009b, p. 361), terms that share a common etymology that could be more appropriately translated into English as 'land-as-soil and land-as-territory' which highlights that land does not merely refer to agriculture but also to the resources of the subsoil, and most importantly, to 'the articulation between the nation-state and its territory' (Brenner and Elden, 2009b, p. 362). For Lefebvre, territory and state are mutually constitutive; his notion of State Space functions as a synonym of territory (Brenner and Elden, 2009b, p. 365). State Space is understood as land and as a historically specific 'political *form* of space produced by and associated with the modern state' understandable 'only through its relation to the state and processes of statecraft'; therefore there can be no state without territory and no territory without a state (Brenner and Elden, 2009b, pp. 362–363). In particular, state territorial strategies mobilise 'state institutions to shape and reshape inherited territorial structurations of political-economic life, including those of state institutions themselves' (Brenner and Elden, 2009b, p. 368). Territory is continually produced and reproduced by state actions, but at the same time, it shapes and conditions the operations and territorial strategies of the state (Brenner and Elden, 2009b, p. 367). Hence, state, space and territory are historically intertwined, and at all scales, State Space is not a static entity transferred through time, it remains in flux, shaped and reshaped through territorial strategies of various kinds (Brenner and Elden, 2009b, p. 370).

As Agnew argued, territory is often taken for granted as a pre-given dimension; this is due to the 'territory effect'. The territory effect naturalises and masks the spatial interventions that are mobilised through interventions of the state (Brenner and Elden, 2009b, p. 373). This naturalisation allows the state to represent its political manipulation of space as either a purely technical intervention or as pre-existing features of the physical realm. It is the state that represents the complexities of 'social spaces of modern capitalism as if they were transparent, self-evident and pre-given' (Brenner and Elden, 2009b, p. 372). Brenner and Elden's reading of Lefebvre provides a path beyond simplistic perceptions of territory by understanding that any State Space, and by extension, any 'territorially configured social space' is the consequence of specific historical forms of economic and political interventions of the state.

Then, State Space strategies, according to Lefebvre's analysis on the politics of space discussed previously, encompass two meanings. First, the struggle and confrontational interaction between the state's – and capital's – attempt to mould space into a rationally manageable abstract entity and the concurrent attempt of social forces to defend, produce and extend spaces for everyday life. Second, the notion of a 'spatial policy' that characterises comprehensive national state systems of spatial management developed by modern capitalist urbanisation (Brenner and Elden, 2009a, p. 367). Furthermore, Brenner and Elden suggest that Lefebvre's analysis of the politics of space focuses not just on the state's spatial strategies, but particularly, on State Space strategies: powerful instruments for the mobilisation of state power to reorganise 'sociospatial relations' (Brenner and Elden, 2009b, p. 368).

The above discussion serves to establish one of the theoretical frameworks of this book: State Space. David Harvey's matrix is helpful as it provides the means to develop a taxonomy of the spatiotemporal categories of each of the documents this book will engage with, as laid out in Table 2.1, in order to deploy Brenner and Elden's reading of Lefebvre as a theorist of State Space as territory. One of the key points of this book is to examine the intersections of spatial policies on the transition towards the Socialist State and PDVSA's extension of its dominant space over non-oil field realms during Hugo Chávez second and third presidential terms. The following section engages with theories on state formation and bureaucracy, to define the manner in which the Venezuelan Petrostate and PDVSA exercise their Bureaucratic Power as parallel State Space(s).

Bureaucratic Power

Before the nineteenth century the promulgation of laws, the economy and assistance for the needy had little to do with the State; the State only became the predominant form to refer to sovereign authority in the latter half of the eighteenth century as it developed into an impersonal and distinct political organisation in charge of the welfare of the people and military defence within a clearly defined territory (Jessop, 1990, pp. 347–348; Miller and Rose, 2008, pp. 55–56). The State as a modern idea involved the distinction between ruler and state apparatus, the State from then on coming to signify the whole of society. For Hobbes, sovereignty lay exclusively on the state, not on the people, and although he defined the state as an impersonal agent, he also affirmed that the state must be represented by a man or an assembly of men whose actions can then be attributed to the state (Walter, 2008, p. 97). By the nineteenth century the state had evolved from limited and centralised apparatuses to an ensemble of 'institutions and procedures of rule over a national territory' (Miller and Rose, 2008, p. 56). The nineteenth century marked the emergence of modern nation states in Latin America after the struggles to gain independence from Spanish colonial rule.

Marxists have traditionally condemned the state as an instrument of domination by a powerful class (Dunleavy and O'Leary, 1987, p. 6). Karl Marx defined the state as an 'economic and political instrument of the dominant class' simultaneously autonomous from the social relations of production and parasitically dependent on them, directly involved in the 'creation and regulation of productive forces' (Brenner and Elden, 2009a, p. 10). This definition has been expanded by Bob Jessop (1990, p. 8) who affirms that to develop a definition of the state it is essential to take into account the complexity of the articulations between state and non-state institutions 'in the overall reproduction of capital accumulation and political domination'. Stressing that the state should be treated as a set of institutions that cannot exercise power, Jessop proposes a definition of the state that encompasses both state institutions and state discourse:

> The core of the state apparatus comprises a distinct ensemble of institutions and organizations whose socially accepted function is to define and enforce collectively binding decisions on the members of a society in the name of their common interest or general will. This broad 'cluster' definition identifies the state in terms of its generic features as a specific form of macro-political organization with a specific type of political orientation; it also establishes clear links between the state and political sphere and, indeed, the wider society.
>
> (Jessop, 1990, pp. 341–342)

With this definition Jesssop aims to emphasise the contradictions in political discourse inherent to any study of the state. He clarifies that the state cannot simply be 'equated with government, law, bureaucracy, a coercive apparatus or another political institution' just as forms of political organisation on the macro level cannot be regarded as state-like, considering that fixed national boundaries do not necessarily determine the emergence of state projects (Jessop, 1990, p. 341). Concerning the idea that the state should be regarded as an ensemble of institutions, Patrick Dunleavy (Dunleavy and O'Leary, 1987, p. 10) proposes two broad categories to define the state: Functional and Organisational. The Functional category, prominent in Marxist approaches to the state, has two strands. One, the state is identified with a range of institutions located outside of the public realm in which any sort of organisation whose objectives overlap with functions of the state become part of the state. Two, the state is defined by its consequences through institutions and patterns of behaviours that stabilise society through social cohesion or social order, thus extending the kind of institutions that can be regarded as part of the state. The Organisational strategy views the state as an ensemble of institutions. The modern state is a particular type of government conformed as a separate set of institutions, with supreme power and sovereignty over its territory and all the individuals within it, clearly differentiated from society creating distinct public and private realms. The state is the authority of law, formulated by state bureaucrats and backed by its monopoly of force, with the capacity to finance its activities through the taxation of its citizens. Dunleavy warns that the set of characteristics of the Organisational definition cannot be applied to all modern countries equally, because modern state structures have evolved and developed differently in different societies influenced by their particular historical circumstances. This applies to modern statecraft in Venezuela, which suffers from the diminished capacity that characterises most rentier states whose main income comes from oil extraction with very little taxation.

Nonetheless, Michel Foucault rejected any attempt to develop a general theory of the state as he believed that the state is neither a universal nor an autonomous source of power but a 'mythical abstraction' granted a place within the field of government (Jessop, 2007, p. 36). Foucault developed an alternative analytic of political power called Governmentality, a new term he derived from a play on the word government. Governmentality focuses on the rationalities that determine practices of government, bringing attention to the mechanisms used to 'know and govern the wealth, health and happiness of populations' (Miller and Rose, 2008, p. 54; Walter, 2008, p. 98). State formation from the fifteenth and sixteenth centuries onwards has been characterised by the 'governmentalization' of the state (Foucault, 1991; Miller and Rose, 2008; Walter, 2008; Joyce and Bennett, 2010), a process that refers to the complex clustering of power systems at the level of the state; it is a process that the state may drive but often neither controls nor authors. This form of power is linked to the proliferation of a wide range of apparatuses related to government, the means to practice it and the nature of those over whom power is exercised.

In this line of thought, Jessop posits that the state does not exercise power; it is an institutional ensemble as the power of the state is 'always conditional and relational' (Jessop, 1990, p. 367). The scope of power rests on 'the action, reaction and interaction of specific social forces' situated inside and beyond of the institutional ensemble as 'it is not the state which acts: it is always specific sets of politicians and state officials located in specific parts of the state system' (Jessop, 1990, p. 367). Similarly, Miller and Rose (2008, p. 10) coincide with Jessop in arguing that the state does not and cannot exercise power; it can only do so through the complex network of organisations, institutions and apparatuses that compose it (2008, pp. 55–56). By the same token, Bennet and Joyce affirm that the state 'rather than a site from which this form of power originates or at which it terminates' is the site where Bureaucratic Power congregates (Joyce and Bennett, 2010, p. 2). Lefebvre (Brenner and Elden, 2009a, p. 12) is critical of this perspective, claiming that it dilutes the concept of power, scattering it in every place and in every single form of subordination, neglecting that the 'real' seat of power lies in the state, in its institutions and constitutions. Nonetheless, bureaucratic forms of organisation and the powers they summon and elaborate take form in many fields, from business to the military; these powers then travel to and from the state, clustering, then redeployed and multiplied. Therefore, when referring to the power of the state, it is more accurate to talk about Bureaucratic Power. Institutions, procedures, strategies and tactics allow the state to exercise power over the population through a bureaucratic apparatus.

To further elaborate on the mechanisms that explain the simultaneous process of fragmentation and centralisation of the Bureaucratic Power of the state, it is useful to integrate the notions of governance conceptualised by R. A. W. Rhodes (2003) and the Shadow State defined by Jennifer Wolch (1990). R. A. W. Rhodes' conceptualisation of different modes of governance and the ideas of hollowing out the state are useful to understand the process of fragmentation and erosion of the Bureaucratic Power of the Venezuelan Petrostate during the transition towards the Socialist State advanced by Hugo Chávez. For Rhodes, Governance is not to be used as a synonym for government, instead governance 'signifies a change in the meaning of government, referring to a *new* process of governing; or a changed condition of ordered rule; or the new method by which society is governed' (Rhodes, 2003, p. 46). Rhodes identifies six different uses of governance: as the minimal state, as corporate governance, as The New Public Management, as 'good governance', as socio-cybernetic systems and as self-organising networks (pp. 46–47).

Governance as the minimal state refers to the redefinition of the 'extent and form of public intervention and the use of markets or quasi-markets to deliver "public" services' in which the size of government is reduced by 'privatization and cuts in the civil service' (Rhodes, 2003, p. 47). Governance as corporate governance defines 'the system by which organizations are directed and controlled' mainly concerned with giving comprehensive direction to the enterprise, by controlling the executive actions of management and complying with expectations of accountability and regulation by the interests that lie outside the boundaries of the corporation (Rhodes, 2003, p. 48). Governance as The New Public Management follows from corporate governance in which government is 'steering' action (policy decisions) by structuring the market to take over 'rowing' functions (delivery of services) (p. 48). Good Governance is a term advanced by the World Bank to shape its lending policies to Third World countries which merges 'the new public management with advocacy to liberal democracy' (pp. 49–50). Governance as a socio-cybernetic system refers to a 'centre-less society' with a polycentric state in which government is no longer the single sovereign authority. The boundaries between public, private and voluntary sectors are blurred by interdependence and shared goals among the social-political-administrative actors in which none holds the monopoly over expertise of information. Here, 'governance is the result of interactive social-political forms of governing' (pp. 50–51). In Governance as self-organising networks, governance is about managing networks. The term network is used to describe the 'several interdependent actors involved in delivering services' (p. 51) in which public, private and voluntary organisations interact to deliver public services. Networks are self-organising, autonomous and self-governing; they resist steering to develop their own policies; in this sense, says Rhodes, they are an alternative to market and hierarchies (pp. 51–52).

Rhodes synthesises that the meaning of Governance 'refers to self-organising, interorganisational networks', a definition that incorporates elements of all the uses enumerated above, with the predominance of the minimal state, socio-cybernetic system and self-organising networks. He lists four shared characteristics of governance: interdependence between organisations, continuing interactions between network members, game-like interactions rooted in trust and regulated by the rules of the game agreed by the participants and a significant degree of autonomy from the state (Rhodes, 2003, p. 53). The state becomes a collection of inter-organisational networks made up of governmental and societal actors with no sovereign actor able to steer or regulate them (p. 57). In summary, Governance blurs the distinction between state and civil society. Rhodes developed his definitions of governance to explain the reforms and changes to British government since the 1980s, which he summarised as 'hollowing out the state' (p. 53). Hollowing out the state is characterised by the downsizing and fragmentation of the public sector through the 'loss of function by central and local government departments to alternative delivery systems' (pp. 53–54). The process of hollowing out leads to fragmentation which diminishes the state's capacity to steer and erodes accountability, in which ultimately the central state loses its grip on the rest of its institutional and bureaucratic apparatus.

In the same vein of characterising the reforms and fragmentation of the state and governance, Wolch focuses on the voluntary sector to identify the rise of the Shadow State, defined as:

> a para-state apparatus comprised of multiple voluntary sector organisations, administered outside of traditional democratic politics and charged with major collective service responsibilities previously shouldered by the public sector, yet remaining within the purvey of state control.
>
> (Wolch, 1990, p. xvi)

As a para-state apparatus the Shadow State undertakes many of the functions of the welfare state, in activities that though not formally part of the state, are 'enabled, regulated, and subsidised' by the state (Wolch, 1990, p. 41). Wolch traces the transformation of the voluntary sector into a Shadow State apparatus to transformations and changes in the welfare state, made possible 'because of the long-standing institutional interdependence of voluntary organisations and the state, which both enables and constrains voluntary action' (p. 15). As state funding had become increasingly important for voluntary organisations, so did the degree of penetration of the state into the organisation, management and goals of voluntary group organisations (p. 15); and as the Shadow State grows, says Wolch, it develops two 'contradictory faces: One represents increased state penetration of many aspects of daily life; the other represents a revitalized democracy in state affairs' (p. 4). While Rhodes used the notion of 'hollowing out the state' to describe the changes in British government due to privatisation and its loss of functions to European institutions and Wolch looked in particular at voluntary organisations in the United Kingdom and the United States, in the case of Venezuela, and more specifically the case of the shift in the relationship between the Petrostate and PDVSA enforced by Hugo Chávez, the process became the opposite of a 'hollowing out of the state', the Petrostate became further inflated, while the ensuing inefficiency of the public sector coalesced in the transfer to PDVSA of many of the new government programs and many functions of the state such as the provision of social housing, food distribution and urban regeneration.

What is useful about this discussion is that by viewing the state as a site where Bureaucratic Power congregates, it provides the ideal framework to understand the process of simultaneous consolidation of PDVSA as a parallel state and the centralisation of Bureaucratic Power in the sole figure of president Hugo Chávez who exercised Bureaucratic Power dramaturgically as the New Magical State, drawing on the definition coined by Fernando Coronil (1997) discussed in Chapter 1.

The adoption of the Bureaucratic Power perspective integrates the idea of the state as an 'institution of territorial governance with vast powers over the material wellbeing of its people' (Mukerji, 2010, p. 82) and the modern state as the only agent with the capacity to manage society and territory on a large scale (Brenner and Elden, 2009a, p. 20). According to Lefebvre, only the state possesses exceptional capacities to funnel long term and large-scale investments in the built environment as well as the sovereign capability to regulate and plan the uses of such investments (Brenner and Elden, 2009a, p. 20). Hence, State Space encompasses far more than territory (Lefebvre, 2003; Brenner and Elden, 2009a, pp. 20–21) given that State Spatial Strategies also shape how industrial development, land use, energy production, transportation and communication reproduce inside and beyond the territory. The Bureaucratic Power perspective is instrumental in understanding how oil wealth permeates all instances of the state, and how the reform of the state apparatus carried out by Hugo Chávez serves to explore how the interfaces between oil, space and culture manifest within a Petrostate. This chapter now turns to explore rentier state theory in order to develop a characterisation of the Bureaucratic Power of the Venezuelan Petrostate.

Petrostate: rentier state theory

The first studies aimed at theorising the impact of externally generated revenues from the exploitation of crude oil on statecraft were mostly focused on oil exporting countries in the Middle East. The concept of the 'rentier state' was coined by Iranian economist Hossein Mahdavy (Beblawi, 1987, p. 51) to refer to states whose main source of income relies on external rent. Rentier states can be traced back to the Spanish Empire and the exploitation of the vast mineral resources found in the Americas.

Nineteenth century economist David Ricardo is credited with introducing the term rent and developing the first comprehensive analysis of resource rent. Resource rent should not be confused with contract rent. The term resource rent was originally related to the use of land and agricultural production, a renewable but scarce resource. The variations of yield of different lands would determine the potential of resource rent for each. Rent, as defined by Ricardo, is the compensation paid by the farmer to the owner of the land for the use of the original and indestructible properties of the soil (Ricardo, 1821, para. 2.2). The payment of rent is also determined by, and varied according to, the investments made in infrastructure by the landlord.

Ricardo extended this principle to the rent of stone quarries and coal mines, in which the compensation is paid to the value of the stone or coal removed from the soil with no connection with the 'original and indestructible powers of the land' (Ricardo, 1821, para. 2.2). From quarries and coal mines this has been extended to oil, gas, diamonds and other mineral resources extracted from the subsoil. In sum, resource rent in Ricardian terms refers to the economic return accumulated for its use in production and the payment to landlords for the right to access and exploit these resources. Thad Dunning in his book *Crude Democracy: Natural Resource Wealth and Political Regimes* speaks of oil rents in Ricardian terms, defined as 'the economic return to natural resource extraction that exceeds labor and other production costs as well as transport costs and some "normal" return to capital' (2008, p. 39), accrued by the state as landlord in exchange to access to resources in the subsoil.

The term rentier state defines a particular relationship between the state and the economy. The rise of the Age of Oil by the late nineteenth century reshaped the landscape of the global economy, with new world powers emerging in the course of the twentieth century. Vladimir Lenin addressed rentier states in his book *Imperialism, the Highest Stage of Capitalism* published in 1917. Lenin regarded the rentier state as the decay of imperialist capitalism in a world 'divided into a handful of usurer states and a vast majority of debtor states' (1978, p. 95). The rentier states he referred to were Great Britain, France, Germany, Belgium, Switzerland and the United States: all industrial states that had become creditor countries who by way of granting loans to politically dependent countries, or former colonies, had deepened their dependency by turning them into debtor states. For Lenin, the rentier state is a 'state of parasitic, decaying capitalism, and this circumstance cannot fail to influence all the socio-political conditions of the countries concerned' (1978, p. 96). Nonetheless, these 'debtor states' could also be rentier estates, since the essence of a rentier state according to economist Giacomo Luciani lies on the 'origin of state revenue, not necessarily in its rent-like character' (1987, p. 13). But to focus exclusively on the state as an entity independent from the economy is not only restrictive but reveals little about the economy producing the state's income.

Rentier states share three distinctive features: rent predominates as a source of income but it's not the only one, the origin of the rent is external, and the majority of the population is the beneficiary of the distribution of the rent generated by a minority engaged in its production (Beblawi and Luciani, 1987, p. 12). For this reason, it is more accurate to speak of a 'rentier economy' in which rent plays a major role in the economy, determining that a rentier economy is more likely to generate a rentier state, both inextricably connected to the emergence of a rentier mentality that shapes politics and development policies (Beblawi and Luciani, 1987, p. 12).

The Petrostate is a particular form of rentier state, in which the majority of the state's revenue comes as oil rent money through exports of oil. Oil rents 'have a strong and decisive influence on the nature of the state' (Luciani, 1987, p. 68), for this reason it is important to look closely at the nature and origin of the state's income:

> It is oil exports that play an essential role in this respect even more than oil production per se: the state in a country in which a lot of oil is produced but none exported may or may not be called rentier, but does not appear to be essentially different from any other state whose income depends on domestic sources. (...) If oil is mostly exported and the income of the state is mostly linked to the exportation of oil, then that state is freed from its domestic economic base and sustained by the economic base of the countries which are importing its oil.
>
> (Luciani, 1987, p. 69)

Luciani introduces two categories of state according to the origin of its revenues: Esoteric States, mostly based on domestic revenue and taxation and Exoteric States, whose revenue mostly originates from abroad. He develops a parallel categorisation based on the predominant function the state plays in regards to its revenues: the Allocation State and the Production State. Luciani clarifies that all states perform some form of allocation, the difference is that in the Production State the allocations are limited by the extent in which the domestic economy 'provides the income which is needed to do so' whereas the Allocation State predominantly distributes 'the income that it receives from the rest of the world' (Luciani, 1987, p. 70). A Production State is correlated to an Esoteric State, while the Allocation State is predominantly an Exoteric State. Drawing on these definitions, the Petrostate is a predominantly exoteric allocative state in which 'revenue derives predominantly (more than 40 per cent) from oil or other foreign sources and whose expenditure is a substantial share of GDP' (p. 70). This establishes the characteristics of the rentier state, especially of oil rentier states like Venezuela, primordial to understand the effects of oil rentierism on statecraft. A rentier economy not only determines the emergence of a rentier state, but also shapes its politics, policy making and the relationship between the state and society. From this point onwards this chapter will refer exclusively to the Petrostate, the particular type of rentier state that concerns this study.

A Petrostate predominantly spends and does not tax, commonly seen as benefitting everyone. Its expenditures, however, are unevenly distributed and irrelevant for political life as there is no incentive to reform political institutions because the abundance of rent money allows it to increase public expenditure to ward off political conflicts (Luciani, 1987, pp. 7, 70). Furthermore, expenditure policies tend to favour the elites, leaving little room for the interests of under-represented sectors since oil revenues act as a buffer against political risks by keeping taxation low and using patronage to buy consensus (Luciani, 1987; Weyland, 2009). This allows political leaders to avoid being 'forced to diffuse political power through representative and democratic institutions' (Dunning, 2008, p. 53) since the collection of oil rents does not require representative pacts between state and citizens. Thus, the main priority of a Petrostate is to extract 'the maximum potential revenue from the rest of the world' (Luciani, 1987, p. 71).

The predominantly distributive nature of the Petrostate is considered a fundamental factor in the tendency towards authoritarianism among most oil producing countries. However, scholars such as Thad Dunning (2008) and Tim Mitchell (2011) challenge the direct correlation between oil and non-democratic governments, but as Terry Lynn Karl (1997), Peter Maass (2009) and Michael Ross (2012) describe, Petrostates do tend to suffer from diminished state capacity especially during times of oil booms. Dunning zooms in on the negative relationship between oil rents and taxation. He argues that the tendency towards authoritarianism or democracy, or the overall economic orientation of a government (whether liberal or interventionist) is independent from the factors that foster the emergence of a Petrostate (Dunning, 2008, p. 37). The key to decipher the commonalities between Petrostates lies in the 'presence of rent-producing natural resources' which shape the fiscal foundation of the state in similar ways across a very diverse set of countries (p. 37). To explain the political effects of oil rents, Dunning establishes a conceptual simplification of the relationship the Petrostate establishes with oil rents:

> it is therefore useful to think of rents as flowing more or less like 'manna from heaven' into the fiscal coffers of the state, even though this is a radical simplification of the actual process by which the state captures rents.
>
> (p. 45)

Mitchell's *Carbon Democracy: Political Power in the Age of Oil* (2011) provides a historical account of the process through which the global networks of the carbon industry in the nineteenth century fostered what has been called the Age of Empire and the Age of Democratisation, and how the rise of coal created democracy in some regions and colonial domination in others (Mitchell, 2011, p. 18). The development of the oil industry, built over the assembly of flows of steam and carbon deepened this difference. Therefore, Mitchell argues that the relationship between the Petrostate and oil rents is far more complex, attention cannot be restricted to the flow of oil money but has to expand to include the process of the production and distribution of oil in both domestic and global networks since the political effects of oil are the outcome of 'particular ways of engineering political relations out of flows of money' (p. 5). Oil rents tend to displace other sources of income as the foundation of public finance, not only because they represent an 'externally generated windfall' but also because the revenue generated from oil is less costly and complex than taxation (Dunning, 2008, pp. 45–46); abundant rents diminish the need to tax citizens, and where taxation is present it tends to be lower. The oil industry becomes the predominant fiscal basis of the Petrostate for both developed and underdeveloped countries (Dunning, 2008, p. 49), magnified by the boom and bust cycles of the oil market (Karl, 1997, pp. 44–52) which alter fiscal and bureaucratic structures.

An account of the political effects of oil is not complete without addressing how the boom and bust cycles of the global oil market influence and thwart state policy making, particularly in Venezuela (Karl, 1997, pp. 5–9). Up until the 1950s it was believed that oil wealth was exceptionally beneficial; that it would provide oil producing countries with the necessary revenues to invest in development (Ross, 2012, p. 2,6). However, the international supply of oil was controlled by a small number of global oil corporations called the Seven Sisters: Anglo-Persian Oil Company (now BP); Gulf Oil, Standard Oil of California (now Chevron), Texaco (later merged with Chevron); Royal Dutch Shell; Standard Oil of New Jersey (Esso/Exxon) and the Standard Oil Company of New York (Mobil now part of ExxonMobil). By the 1960s and 1970s most oil producing countries in the Global South embarked on a wave of nationalisation of their oil industries, setting up their own state-owned oil companies; this period also saw the rise of the Organisation of Petroleum Producing Countries, OPEC. OPEC radically transformed the international oil markets and the scale and volatility of oil revenues from then onwards, creating a number of political, economic and social distortions (Ross, 2012, p. 7) that triggered the emergence of a new phenomenon: the oil curse.

The peculiar properties of oil rent and its effects on the politics and economy of a Petrostate are the cornerstone of the oil curse (Ross, 2012, p. 5). The oil curse determines that economies dependent on natural resource extraction are far more likely to endure chronic political and economic distortions (Corrales and Penfold, 2011, p. 71; Ross, 2012, p. 12). This is due to the nature of oil rents: 'the revenues it bestows on governments are unusually large, do not come from taxes, fluctuate unpredictably, and can be easily hidden' (Ross, 2012, p. 6). Javier Corrales and Michael Penfold's study *Dragon in the Tropics: Hugo Chávez and the Political Economy of Revolution in Venezuela* (2011) addresses the political legacy of Hugo Chávez to re-examine the oil curse, affirming that a direct correlation between fluctuating oil prices and diminished state capacity is an oversimplification because 'the impact of oil in economics and politics often depend on institutional factors' (Corrales and Penfold, 2011, p. 72). They expand on the oil curse to propose the 'institutional resource curse' to describe that oil flows through, and interacts with, a series of pre-existing institutional configurations that shape the effects of oil at times of boom or bust (pp. 72–75). Whether a resource curse materialises is dependent on the interactions between the existing institutional apparatus and oil rents. Therefore, they argue that the rise to power of Hugo Chávez and his Bolivarian revolution can only be understood by addressing the institutional resource curse perspective (p. 72): 'it was not the price of oil alone that helped the regime consolidate power; it was the institutional changes that were made prior to the oil boom together with the oil boom that did the trick' (Corrales and Penfold, 2011, p. 90), changes traced in detail in Chapter 1, such as the new Constitution, the expansion of PDVSA's scope of activities to non-oil related areas and the reform of the Organic Law of Hydrocarbons among a myriad of changes and reforms to the state apparatus to transition towards the Socialist State.

Hugo Chávez's Petro-Socialism was the higher stage of his anti-neoliberal political platform, putting in place social policies 'that clearly contradicted neoliberal orthodoxy' (Parker, 2006, p. 70). It is important at this point to clarify the intersection between oil capitalism and Neoliberalism, to address the contradictions inherent to Petro-Socialism. Tim Mitchell argues that oil not only played a key role in shaping the global economy but 'it also shaped the project that would challenge it, and later provide a rival method of governing democratic politics: the "market" of neoliberalism' (Mitchell, 2011, p. 141). Mitchell argues that neoliberalism was envisioned by Hayek 'as an alternative project to defeat the threat of the left and of populist democracy' (p. 141). David Harvey defines Neoliberalism as 'the elevation of market-based principles and techniques of evaluation to the level of state-endorsed norms' in which the role of the state is to provide the appropriate institutional frameworks to liberate entrepreneurial freedoms, free markets and free trade; the authority of the state relies heavily on economics (Harvey, 2005, pp. 2, 6–7). Nonetheless, says Harvey, neoliberal theory dictates that the state should not venture and its interventions on markets should be kept to a minimum since the state cannot 'possess enough information to second-guess market signals (prices) and because powerful interest groups will inevitably distort and bias state interventions (particularly in democracies) for their own benefit' (p. 7). As neoliberal theorists are suspicious of democracy, they tend to 'favour governance by experts and elites' preferring governments by executive order over 'democratic and parliamentary decision making' (p. 66). Thus, marking a shift from government as 'state power on its own' to governance as 'a broader configuration of state and key elements of civil society' (p. 77). Conversely, Colin Crouch (2011, p. vii) argues that the financial collapse of 2008–2009 that involved the world's leading banks challenged the idea that the free market provides the best means to satisfy human aspirations over the state and politics. He also disputes the claim that neoliberalism is devoted to the free market by affirming that, in practice, it commits to 'the dominance of public

life by the giant corporation' (p. viii); he further elaborates that the political conflicts that stem from the confrontations between the state and the market mask
the existence of this 'third force' which is 'more potent than either and transforms the workings of both'. Although the Venezuelan Petrostate of the Chávez
era cannot be characterised as neoliberal, Harvey's statement that the boundaries
between state and corporate power have become very porous can certainly be
applied here (Harvey, 2005, pp. 77–78). This theorisation is helpful to highlight
the shift in the relationship between Hugo Chávez's anti-neoliberal politics and
the role of PDVSA. Chávez publicly criticised PDVSA's corporate practices, he
discursively reframed PDVSA as a 'socialist company' at the service of his
revolution, and once the oil company was regarded as 'revolutionary' he surren-
dered many of the bureaucratic powers of the state to the state-owned oil
company. Hugo Chávez, aware of the many deficiencies and diminished capacity of the state, believed that 'an oil company would succeed where government
ministries might not' (Maass, 2009, p. 215). This process had ramifications in
the bureaucratic de-professionalisation of PDVSA and the reshaping and recon-
stitution of State Space in the advancement towards a new socialist order, con-
firming that a state that controls the production of oil energy also controls the
production of State Space. Thus, there is a direct relationship between the State
Space of the Petrostate, flows of oil rent and the composition of its Bureaucratic
Power. At the centre of this relationship lies the city, in particular the capital
metropolis, as it is the 'material space where human and political actions manifest' (Lefebvre, 2003, p. 224). Therefore, it is important to understand the mechanisms and the processes under which cities have emerged, as well as the role oil
wealth and Bureaucratic Power have played in urban development and the production of urban society and culture in the twentieth century, particularly in contemporary Venezuela.

Culture as a Resource in and within the Petrostate

The city is the centre of the material effects and the physical space of the State
(Lefebvre, 2003, pp. 252–253). The most evident transformations brought by oil
capital in oil producing countries such as Venezuela are reflected in large infrastructures, shifts in patterns of consumption, social relations, rapid urbanisation
and negotiations between states and oil corporations to secure revenue and
capital accumulation. Oil wealth affects the social order in very tangible ways
through the radical transformation of the built environment. In unpacking the
relationship between oil and the state, its material effects on urban development
need to be addressed in order to understand the changes in society and culture.

It wasn't until the Industrial Revolution that the city began to be considered a
unique form of social life; it is for this reason that urban studies is a fairly recent
discipline when compared to the ancient history of the city (Sennett, 1969, p. 3).
For Marx and Engels what separates the urban from the rural is the division and
specialisation of physical and spiritual labour. They equated the change from rural
to urban with a shift from barbarity to civilisation; a shift that remained constant in

the history of civilisation (Marx and Engels, 1974, p. 55). Rural and urban are also differentiated in terms of the source of their wealth: land property for the former while for the latter, capital developed independently from territorial property based on labour and economic exchange. Moreover, the city is above all the domain of intellectual labour, from there all matters are governed, economic and ideological power emanates, and the moral and religious fate of the nation is decided. The city is also where large crowds gather, it is a place of pleasure and satisfaction of needs; it is also the domain where a materialistic interdependence is developed constrained to alienated labour, inciting consumption through the multiplication of products (Marx and Engels, 1974, p. 56; Remy and Voye, 1976, pp. 245–251). The division of labour modified the social organisation of the city creating a new class structure, the bourgeoisie, born in the city, separate from the proletariat comprised of workers, guilds and plebes. The internal differentiation of the bourgeoisie determined, in turn, the differentiation and hierarchy between cities and the prevalence of all cities over the countryside.

Urban Sociology emerged as a sub-discipline of Sociology in the twentieth century, with an interest in studying the city and its relationship with society and inequality, the nature of urbanism as well as the role of the state in urban development and the distribution of urban resources (Stevenson, 2013, pp. 9–10). Early urban sociologists were mainly concerned with formulating an all-encompassing theory that could explain all cities, regardless of their histories, geographies and cultures (p. 10). These overarching concepts and theories were drawn mainly from the study of Western cities, specifically those of the Global North.

The Chicago School of Urban Sociology was a major influence in urban studies for most of the twentieth century. Preoccupied with developing a coherent conceptualisation of the city and urban life, The Chicago School defined the city as a biological entity that 'adapted systematically and predictably in response to changes in population, demography and the physical environment' (Stevenson, 2013, p. 11). It believed that urbanisation as well as technological, social and cultural processes were inseparable from modernisation; it was also concerned with the 'subjective factors' that determined city morphology such as the location of particular ethnic groups in certain neighbourhoods, studying residents' personal tastes and social status, the way urban space reflected the residents' characteristics of its inhabitants, or caused or amplified social problems such as violence and crime. The Chicago School believed that empirical research was crucial to understand the city and urban culture, establishing both the conceptual parameters and methodological approach that would greatly influence urban research over the twentieth century (p. 12). According to The Chicago School the only function of the state was to support 'social and urban needs', but it never questioned its role in the processes of urbanisation and urbanism (p. 11).

The dominance of The Chicago School was challenged in the 1970s by Marxists, who criticised its inability to predict and explain the urban unrests and social crises of the 1960s. Marxists viewed the city as a site of conflict; they affirmed that capital shaped the spatial structures of cities and was the fundamental cause of social inequality; they emphasised the role of capital accumulation and class struggle in the process of urbanisation and the allocation of urban resources (Stevenson, 2003, pp. 35–36, 2013, pp. 12–13). The Marxist perspective marked the emergence of a new urban sociology dominated by two approaches: one conceptualised the city as 'a system comprised of interdependent networks and components' and a 'localization of social forces', while the other asserted that capitalism 'created cities that are sites of inequality and function to protect capitalism, private property and the accumulation of wealth' (Stevenson, 2013, pp. 10–13). These two approaches were divided along national lines. On one side, French Marxists were concerned with 'how space is used in the process of social reproduction' and the city as the site of social segregation and inequitable distribution of urban resources. On the other side were American urban studies, concerned with studying cities as channels for the circulation of capital and labour discipline.

This study adopts the Marxist approach to urban studies which focuses on the way power, politics and economics manifest in the city as well as on the role global processes and the state play in supporting capitalism through the management of urban populations and urban space (Stevenson, 2013, p. 13). Under the influence of French Marxist philosopher and sociologist Henri Lefebvre, Marxist urban sociologists distinguished between the 'economically determined organisation of urban space' and urbanism 'as a cultural form of collective self-expression' (Zukin, 2006, p. 105). They also viewed the city as a site of conflict where capital shapes the spatial structures of cities (Stevenson, 2003, pp. 35–36, 2013, pp. 12–13). Henri Lefebvre divided the history of cities into three consecutive eras: the agrarian era, the industrial era and the urban era (Lefebvre, 1976, p. 65). The urban era had just begun, as it was waiting to be explored and constructed, manifested in the process of city dispersion encompassing social practices, symbols and works of culture. Therefore, for Lefebvre, urban societies were still in the process of formation. This understanding veers the discussion towards an 'epistemology of the urban' that functions as a wider theoretical framework to study the process of total urbanisation of modern societies (Stanek, 2011, p. xiii).

Lefebvre observed that the world was headed towards complete urbanisation, a process deeply enmeshed with the survival of capitalism (Lefebvre, 1991, p. 4; Stevenson, 2013, p. 15). As the capitalist industrial economy grows, the city physically extends its boundaries disseminating its 'urban fabric' through the materialisation of suburbs, industrial complexes, satellite cities; this process of urban*isation* gradually absorbed or colonised the countryside and its agrarian mode of production: 'a vacation home, a highway, a supermarket in the countryside are all part of the urban fabric' (Lefebvre, 1991, p. 4). For Lefebvre the city designates a concrete and clearly defined *object*, a built entity, however diffuse its boundaries might be, while *urban* is an abstract concept, a social practice inherent to the capitalist mode of production emancipated from the materiality of the city and a carrier of modernity and culture (Lefebvre, 1976, p. 68). The urban then is both spatial and temporal: it extends through space and develops over time. In this sense, Manuel Castells affirms that no urban world exists outside of modern capitalism, defining urbanism as the culture of the city of 'industrial capitalism and the capitalist state' (Zukin, 2006, p. 105; Stevenson, 2013, p. 14). Along the same lines, David Harvey distinguishes between 'the city as built form' and the urban 'as a way of life', concurring with Lefebvre's definitions. Harvey views cities as 'landscapes of power' (Zukin, 2006, pp. 105–106) in which capital has the power to make and remake the urban, opening paths for uneven or contradictory development across cities and across the globe.

The term urban, understood as the immaterial carrier of modernity and culture, will be used in this book to refer to the social, political and cultural practices of industrialised capitalist societies. City will be used to refer to the built form (buildings, infrastructure, roads, networks, etc.) and the geographical expansion it occupies, where urban societies settle and conglomerate. Hence, the term 'urban space' categorises specific and clearly delineated areas of congruence between urban and city.

While Lefebvre and Harvey defined the urban as the carrier of the modern ways of life and culture shaped by capitalism, Sharon Zukin defines culture in urban terms as the 'habits carried through space and time, refined through the interaction with church and state, and asserted as a means of differentiation and independence' (Zukin, 1995, p. 263). She also argues that culture is 'what cities do best' (p. 264) as it also refers to the idea of a collective urban lifestyle, a meaningful if somewhat 'conflictual source of representation' that encompasses a diverse range of ethnicities, lifestyles and images. The city is where cultural industries are located, where artists and performers produce their work, but most importantly, as a built form it contains a visual repertoire that becomes a public language that makes visible the implicit values of a particular group (p. 264). How a city looks and feels is the reflection of the uses of aesthetic power and concepts of order and disorder used to dictate what and who should be visible and where.

For Zukin, culture can be understood as the legitimate ways in which a particular group creates its own niche in society, encompassing habits carried through time and space, distilled through interactions with the state (Zukin, 1995, p. 263). Culture can also function as an instrument of spatial and social stratification in which every space of the city caters its visual consumption to a specific constituency (p. 36). Urban space is produced in synergy with 'capital investment and cultural meaning' (pp. 23–24). Elites reconfigure urban space imposing their point of view; any change in the reorganisation of urban space manifests in changes of the visual representation of the city. For this reason, it is imprecise to talk about urban culture as universal and homogeneous, rather one should ask 'which culture?, which cities?' (p. 264).

The combination of 'traditional economic factors of land, labor and capital' (Zukin, 1995, p. 7) determine the built environment of the city as well as the manipulation of the symbolic language of exclusion and belonging where culture is often turned into 'a set of marketable images' (p. 265). This commoditisation of culture functions as a symbolic economy, a feature common to all cities, consisting of 'a continual production of symbols and spaces that frames and gives meaning to ethnic competition, racial change and environmental renewal and decay' (p. 265). However, in the symbolic economy the identities of places tend to be established by their visual delights, unifying the 'material practices of finance, labor, art, performance, and design' (p. 9). Culture also functions as an instrument of control, determining, through images, memories and symbols, who belongs to certain places of the city and who doesn't (p. 1). Elites view culture as an instrument to 'restore civility', they devise cultural strategies, magnifying the cultural role of urban institutions to 'reconstruct the meaning of urban spaces to give the appearance of a common public culture'; defined by Zukin as the process of negotiation of images that are commonly accepted by large groups of people in which public space plays a fundamental role, as it is the place where strangers come together and where the boundaries of urban society are negotiated (pp. 10, 270). In this line of thought, beyond the issue of ownership, public space should be understood as any 'shared urban space', whether publicly or privately owned, where large groups of people congregate for a wide range of purposes and where daily encounters with difference takes place (pp. 52–53). The creation of a public culture requires the manipulation of public space for certain kinds of expected social interactions and creates a particular visual representation of the city (p. 24). This creates tensions around cultural politics and the occupation of the city, contradicting the spirit of public culture and evidencing the difficulty of conceiving culture as both an elite resource and a democratic good (p. 270).

Culture then has an intrinsic political value and can be used to serve political purposes; thus capitalism and oil rentierism, as discussed earlier, shape a particular form of making politics within Petrostates within which culture is embedded. Scholarly work exploring the cultural dimensions of oil capitalism in the humanities is a very recent development, it coincides with the crises of the oil economy of the late twentieth century, which generated new interests in understanding how oil became the world's dominant energy commodity in order to develop critical analysis of the symbolic and cultural forms of oil capitalism (Szeman, 2013a, 2013b; Barrett and Worden, 2014, p. xxi; LeMenager, 2014; Lord, 2014). A point of departure to understand the grip of oil politics over culture is *Oil Culture* edited by Ross Barrett and Daniel Worden (2014). *Oil Culture* explores the presence of oil in culture, how oil works within culture and how oil has shaped how we imagine contemporary life. It examines a particular model of culture that designates 'a dynamic field of representations and symbolic practices that have infused, affirmed, and sustained the material armatures of the oil economy' and 'the particular modes of everyday life that have developed around oil use in North America and Europe since the nineteenth century (and have since become global)' (Barrett and Worden, 2014, p. xxiv). *Oil Culture* is a contribution towards legitimising the field of Oil Studies within the discipline of cultural studies. Barrett and Worden note the absence of literature that 'addresses oil as a cultural material in everyday experience and aesthetics' (p. xx). They set to explore the connections between 'oil capitalism and cultural representation':

As a material whose utility is largely realized through its own destruction, oil requires creative accounts of its worth that depart from its physical form. As a substance that can (at least initially) be extracted without much work, moreover, oil encourages fetishistic representations of its value as magical property detached from labor.

(Barrett and Worden, 2014, p. xxiv)

Oil Culture primarily focuses on the multiple meanings of oil within the cultural imagination of the United States, with some essays examining the cultural dimensions of oil in Europe and other Petrostates such as Canada, Mexico and Niger. Barrett and Worden speak, for example, of a number of literary texts in the United States addressing the cultural and social effects of oil (Barrett and Worden, 2014, p. xxi), and hint that this may be the case in the rest of the world. Actually, Venezuela has a rich production of literary texts and essays that document and portray how oil was intertwined with the presence of foreign dominant forces such as international oil corporations, viewed as mechanisms of domination that threatened national culture and reshaped economic, social and cultural landscapes. In Venezuela, oil not only became the main mode of production and subsistence, it shaped statecraft and society formation and influenced every aspect of national life. Since the early twentieth century, right at the start of its own era of oil, many texts emerged to account and explain the direct social and cultural effects of the oil industry in Venezuela. The first publications to assess oil in cultural terms in Venezuela were produced by the Venezuelan Marxist anthropologist (and former oil camp dweller) Rodolfo Quintero, who published a series of works exclusively devoted to analysing the pervasive social and cultural effects of oil, most of it written around the period leading up to the oil boom and the nationalisation of the oil industry in 1975: the seminal *La cultura del petróleo: ensayo sobre estilos de vida de grupos sociales de Venezuela* (1968), *Antropología del petróleo* (1972), *La cultura nacional y popular: ensayo antropológico sobre aspectos de la dependencia cultural en Venezuela* (1972) and *El petróleo y nuestra sociedad* (1975). Quintero's first work, *La cultura del petróleo* (1968) is particularly relevant. A number of studies on the effects of oil in Venezuela were published on the aftermath of the 1980s oil crash, but most notably in recent years, prompted by Hugo Chávez's entrenchment of the country's dependency on oil. Two notable contributions by Venezuelan scholars are Fernando Coronil's seminal work *The Magical State: Nature, Money and Modernity in Venezuela* (1997) which accounts the ways in which the state's monopoly over the oil economy allowed it to enact collective fantasies of progress by way of spectacular projects, becoming the single agent endowed with the 'magical powers' to transform the nation; and Miguel Tinker Salas' *The Enduring Legacy: Oil and Society in Venezuela* (2009) which explores the oil camps of transnational companies as cultural laboratories that promoted forms of citizenship and ways of life related to the global oil industry. Overall, the literature that assesses the effects of oil on modern social life and thought beyond the realms of politics, economics, technology and energy is a fairly recent development in international academic fields. Frederick Buell's essay *A Short History of Oil Cultures* (2014) develops an approach towards the theorisation of the connection between oil and culture, exploring the close relationship between the development of energy industries and the cultural transformations of the nineteenth and twentieth centuries. Buell conceptualises history as a succession of systems of energy, comprised by economic, technological, environmental and socio-cultural relationships. He argues that oil has become essential to every single aspect of contemporary life, it underlies 'material and symbolic cultures' that partially determines 'cultural production and reproduction' (Buell, 2014, p. 70). Nonetheless, he notes that a gap remains in the conceptual response to the relationship between oil and culture.

Buell's approach towards the cultural effects of oil coincides with Mitchell's (2011) focus on addressing first the mechanics of the development of the coal industry in the nineteenth century. Coal-based mechanical power brought with it the promise of never-ending progress, which inaugurated a double-edged discourse of the effects of fossil energies, oil particularly (Buell, 2014). The consolidation of global oil capitalism in the twentieth century inaugurated a 'new cultural regime' in which modernity engendered extreme exuberance and catastrophe (Buell, 2014, p. 82). Oil's exuberance manifested as modernisation, progress, economic growth, technological development, imperial expansion, rapid urbanisation, sanitation and consumerism. Its inseparable catastrophic side manifests as environmental disasters, air pollution, grim working conditions in the oil fields, colonial domination, economic crises caused by the cycles of boom and bust of oil prices, rural to urban massive migrations and destructive war machinery (tanks, aeroplanes, trucks, submarines). Thus, the consolidation of global oil capitalism is closely related to the rise of modernist culture, as oil exerts a significant but seldom acknowledged influence in contemporary cultural production.

The term culture originates from farming concepts such as cultivation, husbandry and breeding, 'all meaning improvement'; culture is seen 'through the eyes of the "farmers of the human-growing fields" – the managers' (Bauman, 2004, pp. 63–64). Metaphorically what the farmer does with the seed, culture does to the 'incipient human beings by education and training' (p. 63). This evokes the Victorian period's understanding of culture as 'moral betterment and spiritual development' achieved through the contemplation of the best of human creations, where aesthetics articulated differences between what is high culture and what is not (Miller and Yúdice, 2002, p. 1; O'Brien, 2014, p. 2). However, by the mid-twentieth century, an anthropological register was connected to culture (Miller and Yúdice, 2002, p. 1), in which culture is about the 'artefacts and activities associated with a given community's "way of life"' (O'Brien, 2014, p. 2). However, the anthropological register can be 'too expansive for certain analytical and practical purposes' lest it considers the concept of culture as a 'realized signifying system' developed by Raymond Williams referring to 'the practices and institutions that make meaning' where 'symbolic communication is the main purpose and even and end in itself' (McGuigan, 2003, p. 24).

George Yúdice's *The Expediency of Culture: Uses of Culture in the Global Era* (2003) explains the utilisation of Culture as a Resource in the context of neoliberalism and globalisation, as an instrument to aid social and economic development. Culture has acquired to an extent the same status as natural resources as a consequence of the process of globalisation which has accelerated the transformation of all realms of modern life into a resource (Yúdice, 2003, pp. 9, 28). The expedience of Culture as a Resource allows its use for economic, social and political purposes. Concurring with Bauman, he asserts that in both cultural and natural resources 'management is the name of the game' (Yúdice, 2003, p. 2). He argues that the use of Culture as a Resource is not a perversion or a reduction of its symbolic dimension; on the contrary, the expediency of Culture as a Resource is a feature of contemporary life; its transformation into a resource traced to a performative force, a style of social relations generated by diverse organised relations between state institutions and society such as schools, universities, mass media, markets and so on (pp. 47, 60–61). Drawing on Judith Butler, Yúdice characterises performativity 'as an act that "produces which it names" revealing the power of discourse to produce realities through repetition' (p. 47, 58), focused in particular on the institutional preconditions and processes by which culture and its effects are produced.

The activities with respect to the arts, cultural industries, humanities and heritage deployed by government, institutions or corporations are summarised as a cultural policy (Mulcahy, 2006, p. 320). Cultural policy is in essence bureaucratic, it is 'the institutional supports that channel both aesthetic creativity and collective ways of life' functioning as a bridge between the anthropological and aesthetic understandings of culture (Miller and Yúdice, 2002, p. 1). McGuigan revisits Raymond Williams' categorisations of cultural policy as 'display' developed in 1984 (Ahearne, 2009, p. 145), expanding them to draw a distinction between cultural policy as display characterised by 'the instrumentalization of cultural resources for economic and political purposes' and cultural policy proper which 'attends to the proper object of cultural policy' (p. 145). Jeremy Ahearne defines cultural policy as not just a 'predefined object for cultural history, but also a particular "lens" through which cultural history more generally can be approached' (p. 142). Ahearne distinguishes between two broad categories of cultural policy: explicit cultural policy and implicit cultural policy (p. 141). Explicit cultural policy is 'any cultural policy that a government labels as such'; they often work on a definition of culture in terms of the consecrated arts. On the other hand, implicit cultural policy is 'any political strategy that looks to work on the culture of the territory over which it presides (or on that of its adversary)', which implies the unintended cultural side effects of different kinds of policy or government actions that are not labelled as 'cultural' (pp. 143–144). Hence, implicit cultural policy is not constrained to the governmental sphere; the actions of transnational commercial organisations have had significant influence on shaping cultural practices as implicit cultural policy also entails the notion of 'soft power' (p. 144). Implicit cultural policy may be the most influential and important form of cultural policy especially when wielded by powerful transnational commercial organisations such as oil conglomerates whose 'policies' are not as susceptible to government control (p. 146). In this sense, implicit cultural policy can be found in the most unlikely places, such as the Venezuelan Organic Law of Hydrocarbons interpreted and implemented as such by PDVSA La Estancia discussed in Chapter 4.

The issue here is then, with what intentions and to what purposes culture is 'exploited', noting that it is currently close to impossible to find public statements that do not instrumentalise culture to improve social conditions or to foster economic growth (Yúdice, 2003, pp. 10–11). Yúdice affirms that nowadays it is impossible not to turn to Culture as a Resource as it is congruent to the way we now understand nature, affecting the way culture is viewed and produced. Powerful institutions such as the European Union, the World Bank and the Inter-American Development Bank understand culture as a 'crucial sphere of invest-ment' meaning that culture is therefore treated like any resource, as James D. Wolfenshohn, the president of the World Bank, affirmed in his keynote address at the conference *Culture Counts: Financing, Resources, and the Economics of Culture in Sustainable Development in 1999*: 'physical and expressive culture is an undervalued resource in developing countries' (p. 13). This has marked a turn to culture as the prime instrument for investing in civil society as culture would not be funded unless it provides an indirect form of return in the form of 'fiscal incentives, institutional marketing or publicity value, and the conversion of non-market activity to market activity' (pp. 14–15) given that 'cultural institutions and funders are increasingly turning to the measurement of utility because there is no other accepted legitimation for social investment' (p. 16). Thus, while the term expediency refers to the merely politic in regards to self-interest, Yúdice's perfor-mative understanding of the expediency of culture 'focuses on the strategies implied in any invocation of culture, any invention of tradition, in relation to some purpose or goal' which is what makes possible to invoke Culture as a Resource 'for determining the value of an action' (p. 38). For Yúdice, the expediency of Culture as a Resource has become, in practice, the only surviving definition (p. 279) in which cultural policy becomes the institutional medium for its invoca-tion. Yúdice's discussion frames one of the approaches of this book to the work of PDVSA La Estancia, a cultural institution with privileged access to the oil rent to be a founder and manager of culture, needing no negotiations with the govern-ment, and as this book demonstrates, it can even surpass in financial and political power other public institutions whose functions overlap with their cultural work over the city enacting its own mechanisms of oil-based cultural policy.

Bauman argues that the notion of management has been endemic to the concept of culture from the beginning, implying an acceptance of an asymmetri-cal social relation: 'the split between acting and bearing the impact of action, between the managers and the managed, the knowing and the ignorant, the refined and the crude' (Bauman, 2004, p. 64). Furthermore, the relationship between managers and managed is inevitably agonistic and conflictive in which two sides seeking opposite purposes cohabit in a constant mode of conflict (p. 64). Adorno acknowledged that conflict was inevitable, because managers and managed need each other: 'culture suffers damage when it is planned and administrated; if it is left to itself, however, everything cultural threatens not only to lose the possibility of effect, but its very existence as well' (p. 64). There is a very material historical relationship between management and culture when considering state formation and particularly state bureaucracy as the actions of the state shape the understanding of culture, and those aspects of the cultural that appear as natural, fixed and unchanging social facts are subject to governmental structures (O'Brien, 2014, pp. 9–10). Tony Bennett's discussion about the relationship between culture and power draws attention to the institutional circumstances that interpret and regulate culture. Bennett conceptualises culture as 'a particular field of government' (McGuigan, 2003, p. 29), encompassing all power/knowledge relations in society. Hence, culture can be treated as a 'historically specific set of institutionally embedded relations of government' which seeks to transform particular forms of thought and conduct of populations, through instruments of control and power such as 'the social body of forms, techniques, and regimes of aesthetic and intellectual culture' (p. 30). In this vein, culture cannot escape the 'ruses of power and public controversy'; it has always been political, the political grip around culture 'has generally been greatest in societies where the state has played a manifest role in its regulation' (p. 25) as has been the case of the dominance of the Venezuelan Petrostate. In Venezuela the cultural sector has been historically dominated by the institutional action of the Petrostate, brought about by the wave of modernisation fostered by the vast oil rents from oil exploitation from the mid-twentieth century onwards, which enabled the expansion of a powerful state-funded cultural apparatus that opened spaces for autonomous cultural production (Silva-Ferrer, 2014, p. 22). This means that a particular articulation is constructed at the intersection between oil, territory, Bureaucratic Power and culture in the contemporary Venezuelan Petrostate.

This book is particularly interested in deciphering the discursive and institutional mechanisms that enabled the cultural arm of PDVSA, PDVSA La Estancia, to interpret and enact the Organic Law of Hydrocarbons, which is in principle a legal instrument limited to matters of fossil fuel extraction and commercialisation, as a parallel policy instrument of territory and culture. A particular set of institutional circumstances enabled the social and cultural arm of PDVSA to supersede the authority of local government to stretch its scope of action to public art and public space to demonstrate that within the extractive logic of the oil industry, oil, territory and culture become inextricable from each other.

PDVSA La Estancia deploys discursive constructions to perpetuate the oil rentier model, to provide the illusion of inexhaustible oil analysed in Chapters 4 and 5 respectively. These discursive constructions and the institutional circumstances that underpin it coalesce in the notion of Culture as Renewable Oil. PDVSA La Estancia's use of farming language discursively equates culture to 'renewable oil', as if culture was a mineral deposit in the subsoil waiting to be mined, extracted, exploited and distributed by PDVSA. Thus, culture becomes, like the subsoil, the exclusive realm and property of the Petrostate, encompassed within PDVSA's State Space. While Hugo Chávez established his political platform as an alternative to neoliberalism, oil capitalism provided the financial resources to advance and implement Petro-Socialism. The cultural arm of the state-owned oil company PDVSA, as an agent of the Bolivarian revolution, understands culture as one of the mediums and instruments to transcend the ills of capitalism through Petro-Socialism.

Conclusion

This chapter outlined the three theoretical premises that guide this book's investigation. First, Brenner and Elden's State Space as Territory serves to characterise the Oil Social District as PDVSA's State Space in Chapter 4, which provides the foundation to examine the discursive mechanisms used by PDVSA La Estancia to implement the Organic Law of Hydrocarbons as an implicit cultural policy in Chapter 5. With this purpose, Harvey's triad of spatiotemporality underpins the approach of Chapters 4, 5 and 6, as it enables to identify each particular spatiotemporal dimension of the four inter-related sets of documents this book examines: policy instruments; public speeches of Hugo Chávez, the President of PDVSA and the General Manager of PDVSA La Estancia; and the 23 adverts of PDVSA La Estancia's campaign 'We transform oil into a renewable resource for you'. Second, this book adopts Bennett and Joyce's perspective of the state as the site where Bureaucratic Power congregates, as it is useful in understanding the process of the simultaneous fragmentation of the state apparatus and centralisation of Bureaucratic Power in the sole figure of Hugo Chávez in the transition towards the Socialist State. Finally, this book expands George Yúdice's notion of 'Culture as a Resource' by interrogating the intersections between the state-owned oil company, culture and urban space, to propose the notion of Culture as Renewable Oil, mobilised through the material space of the city, reframed as an oil field conceptualising a symbiotic and cyclical relationship between oil, territory, Bureaucratic Power and culture.

References

Agnew, J. (1994) 'The Territorial Trap: The Geographical Assumptions of International Relations Theory', *Review of International Political Economy*, 1(1), pp. 53–80.

Agnew, J. (2010) 'Still Trapped in Territory?', *Geopolitics*, 15(4), pp. 779–784.

Ahearne, J. (2009) 'Cultural Policy Explicit and Implicit: A Distinction and Some Uses', *International Journal of Cultural Policy*, 15(2), pp. 141–153.

Barrett, R. and Worden, D. (eds) (2014) *Oil Culture*. Minnesota: University of Minnesota Press.

Bauman, Z. (2004) 'Culture and Management', *Parallax*, 10(2), pp. 63–72.

Beblawi, H. (1987) 'The Rentier State in the Arab World', in Beblawi, H. and Luciani, G. (eds) *The Rentier State*. London: Croom Helm, pp. 49–62.

Beblawi, H. and Luciani, G. (1987) 'Introduction', in Beblawi, H. and Luciani, G. (eds) *The Rentier State*. London: Croom Helm, pp. 1–21.

Brenner, N. and Elden, S. (2009a) *Henri Lefebvre. State, Space, World: Selected Essays.* Minneapolis: University of Minnesota Press.

Brenner, N. and Elden, S. (2009b) 'Henri Lefebvre on State, Space, Territory', *International Political Sociology*, 3(4), pp. 353–377.

Buell, F. (2014) 'A Short History of Oil Cultures; or, the Marriage between Exuberance and Catastrophe', in Barrett, R. and Worden, D. (eds) *Oil Culture*. Minnesota: University of Minnesota Press, pp. 69–88.

Coronil, F. (1997) *The Magical State: Nature, Money, and Modernity in Venezuela.* Chicago: University of Chicago Press.

Corrales, J. and Penfold, M. (2011) *Dragon in the Tropics. Hugo Chávez and the Political Economy of Revolution in Venezuela.* Washington D.C.: The Brookings Institution.

Crouch, C. (2011) *The Strange Non-Death of Neoliberalism.* Cambridge, UK: Polity Press.

Dunleavy, P. and O'Leary, B. (1987) *Theories of the State: The Politics of Liberal Democracy.* London: Macmillan Education Ltd.

Dunning, T. (2008) *Crude Democracy. Natural Resource Wealth and Political Regimes.* New York: Cambridge University Press.

Elden, S. (2004) *Understanding Henri Lefebvre.* London: Continuum.

Foucault, M. (1991) *The Foucault Effect. Studies in Governmentality.* Edited by G. Burchell, C. Gordon, and P. Miller. Chicago: The University of Chicago Press.

Harvey, D. (2005) *A Brief History of Neoliberalism.* Oxford, UK: Oxford University Press.

Harvey, D. (2006a) 'Space as a Keyword', in Castree, N. and Gregory, D. (eds) *David Harvey: A Critical Reader.* London: Blackwell Publishing Ltd, pp. 270–293.

Harvey, D. (2006b) *Spaces of Global Capitalism. Towards a Theory of Uneven Geographical Development.* London: Verso.

Jessop, B. (1990) *State Theory: Putting the Capitalist State in its Place.* Cambridge, U. K.: Polity Press.

Jessop, B. (2007) 'From Micro-powers to Governmentality: Foucault's Work on Statehood, State Formation, Statecraft and State Power', *Political Geography*, 26(1), pp. 34–40.

Joyce, P. and Bennett, T. (eds) (2010) *Material Powers: Cultural Studies, History and the Material Turn.* London: Routledge.

Karl, T. L. (1997) *The Paradox of Plenty: Oil Boom and Petro-States.* Berkeley: University of California Press.

Lefebvre, H. (1976) *Espacio y política: el derecho a la ciudad, II.* Barcelona: Península.

Lefebvre, H. (1991) *The Production of Space.* Oxford, UK: Blackwell Publishing Ltd.

Lefebvre, H. (2003) 'Space and the State', in Brenner, N. and Elden, S. (eds) *State/Space. A Reader.* Blackwell Publishing Ltd.

LeMenager, S. (2014) *Living Oil. Petroleum Culture in the American Century.* Oxford, UK: Oxford University Press.

Lenin, V. I. (1978) *Imperialism, the Highest Stage of Capitalism*, 17th edn. Moscow: Progress Publisher.

Lord, B. (2014) *Art & Energy: How Culture Changes.* Chicago: University of Chicago Press.

Luciani, G. (1987) 'Allocation vs. Production States: A Theoretical Framework', in Beblawi, H. and Luciani, G. (eds) *The Rentier State.* London: Croom Helm, pp. 63–78.

Maass, P. (2009) *Crude World.* London: Allen Lane.

Marx, K. and Engels, F. (1974) 'Capítulo 8. La ciudad, la división del trabajo y el surgimiento del capitalismo', in *La Ideología Alemana*. Montevideo: Grijalbo, pp. 55–70.

McGuigan, J. (2003) 'Cultural Policy Studies', in *Critical Cultural Policy Studies: A Reader*. Malden, MA: Blackwell, pp. 23–42.

Miller, P. and Rose, N. (2008) *Governing the Present: Administering Economic, Social and Personal Life*. Cambridge: Polity.

Miller, T. and Yúdice, G. (2002) *Cultural Policy*. London: Sage Publications Ltd.

Mitchell, T. (2011) *Carbon Democracy: Political Power in the Age of Oil*. London: Verso.

Mukerji, C. (2010) 'The Unintended State', in *Material Powers: Cultural Studies, History and the Material Turn*. London: Routledge, pp. 81–101.

Mulcahy, K. (2006) 'Cultural Policy: Definitions and Theoretical Approaches', *The Journal of Arts Management, Law, and Society*, 35(4), pp. 319–330.

O'Brien, D. (2014) *Cultural Policy: Management, Value and Modernity in the Creative Industries*. New York: Routledge.

Parker, D. (2006) 'Chávez and the Search for an Alternative to Neoliberalism', in Ellner, S. and Tinker Salas, M. (eds) *Venezuela: Hugo Chavez and the Decline of an Exceptional Democracy*. Plymouth, England: Rowman & Littlefield Publishers, pp. 60–74.

Remy, J. and Voye, L. (1976) 'Capítulo VII. Karl Marx (1818–1883)', in Remy, J. and Voye, L. (eds.) *La ciudad y la urbanización*. Madrid: Instituto de Estudios de Administración Local.

Rhodes, R. A. W. (2003) *Understanding Governance. Policy Networks, Governance, Reflexivity and Accountability*. London: Open University Press.

Ricardo, D. (1821) 'Chapter 2 – On Rent', in *On the Principles of Political Economy and Taxation*, 3rd edn. London: Empiricus Books, p. 22.

Ross, M. (2012) *The Oil Curse How Petroleum Wealth Shapes the Development of Nations*. Princeton, NJ: Princeton University Press.

Sennett, R. (1969) *Classic Essays on the Culture of Cities*. New York: Appleton-Century-Crofts.

Silva-Ferrer, M. (2014) *El cuerpo dócil de la cultura: poder, cultura y comunicación en la Venezuela de Chávez*. Madrid/Frankfurt am Main: Biblioteca IberoAmericana/Vervuert.

Stanek, L. (2011) *Henri Lefebvre on Space: Architecture, Urban Research and the Production of Theory*. Minneapolis: University of Minnesota Press.

Stevenson, D. (2003) *Cities and Urban Cultures*. Maidenhead: Open University Press.

Stevenson, D. (2013) *The City*. Cambridge, UK: Polity.

Szeman, I. (2013a) 'Crude Aesthetics: The Politics of Oil Documentaries', *Journal of American Studies*, 46(2), pp. 423–439.

Szeman, I. (2013b) 'How to Know about Oil: Energy Epistemologies and Political Futures', *Journal of Canadian Studies*, 47(3), pp. 145–168.

Tinker Salas, M. (2009) *The Enduring Legacy: Oil, Culture, and Society in Venezuela*. Durham: Duke University Press.

Walter, R. (2008) 'Reconciling Foucault and Skinner on the State: The Primacy of Politics?', *History of the Human Sciences*, 21(3), pp. 94–114.

Weyland, K. (2009) 'The Rise of Latin America's Two Lefts: Insights from Rentier State Theory', *Comparative Politics*, 41(2), pp. 145–164.

Wolch, J. (1990) *The Shadow State: Government and Voluntary Sector in Transition*. New York: The Foundation Center.

Yúdice, G. (2003) *The Expediency of Culture: Uses of Culture in the Global Era*. London: Duke University Press.

Zukin, S. (1995) *The Cultures of Cities*. Massachusetts: Blackwell.

Zukin, S. (2006) 'David Harvey On Cities', in Castree, N. and Gregory, D. (eds) *David Harvey: A Critical Reader*. Oxford, UK: Blackwell Publishing Ltd, pp. 102–120.

3 Territory effect, the New Geometry of Power and the construction of a Petro-Socialist State Space

Hugo Chávez was the leader and central figure of the Bolivarian revolution, from his landslide win of the presidential election in 1998 to beyond his death in March 2013, at the start of his fourth term in power, envisioned as the period for the culmination of the transition towards the socialist state and communal popular power. Hugo Chávez embodied a New Magical State (Coronil, 1997; López-Maya, 2007); he condensed the Bureaucratic Power of the state in his persona to completely refashion the country using the 'magical powers' of the oil windfall to push forward a model based on the dyad oil wealth-socialist state, Petro-Socialism, using oil wealth as many governments had before him, as the vehicle for transforming the nation.

His unexpected death was particularly problematic because the transition to a socialist state was ultimately tied to his persona; when he introduced Petro-Socialism to the National Assembly in August 2007 (López-Maya, 2013, p. 102) he emphasised that this was *his* personal project, penned by his own hand. Hence, there are inevitable entanglements between Chávez's discourse of Petro-Socialism and the mobilisation of Bureaucratic Power to reorganise socio-spatial relations through new spatial strategies for the materialisation of a Socialist State Space (Brenner and Elden, 2009). If policy instruments are legal entities of space that conceptualise the administrative boundaries of State Space that function as containers of absolute-representations of space, what are the representations of space produced under Petro-Socialism?

A study of Chávez's national plans for the nation and the new legal instruments for the management of territory elucidates what kind of spatiality could be conceived in Petro-Socialism, informed by a particular institutional resource curse (Corrales and Penfold, 2011) while it entrenches diminished state capacity to enable PDVSA to construct its own parallel State Space. This chapter engages with the underlying contradictions of Petro-Socialism that inform the distortions of the institutional apparatus of urban governance fostered by Chávez's transfer of Bureaucratic Power to PDVSA. It also provides the basis to understand the diverse institutional and discursive mechanisms utilised by PDVSA La Estancia to appropriate and re-imagine the material space of the city as an oil field within the jurisdiction of the Oil Social Districts.

The lack of coherence and inherent contradictions within Petro-Socialism also points to the failure of the consolidation of the New Geometry of Power as a coherent spatial policy for the Socialist State. If, as Brenner and Elden have argued, territory and state are mutually constitutive, the later section of this chapter shows that in the process of re-constituting the territory, a void was created in spatial policies that destabilised the bureaucratic structure of the institutions that govern Caracas, according to the ambition to create a new Socialist State. This is why territory was fundamental for Chávez's political project, although in practice the spatial and territorial strategies of the Socialist State had to coexist uncomfortably with the structures of the Fourth Republic it was meant to substitute. In this context, the New Magical State's territory effect failed to naturalise and mask its spatial interventions.

This chapter is divided into two parts. First, it examines the centrality of the notion of territory for the consolidation of the Socialist State guided by the principle of the New Geometry of Power (the Fourth Engine of the revolution), requiring the dismantlement of the existing institutional apparatus through the abrogation of the Organic Law for the Organisation of Territory of 1983 and the Organic Law of Urban Planning of 1987, the key spatial policy instruments that had shaped the role of institutions with competencies in urban planning; followed by the creation of new policy instruments to reconfigure the national territory as a Socialist State Space. Subsequently, the chapter is concerned with tracing the process of abrogation and substitution of the legal framework of political-administrative territorial management set up in the 1980s, a process fraught with inconsistencies that opened a legal vacuum that diminished State Space authority, enabling PDVSA La Estancia to establish that the Oil Social Districts, defined by the new Organic Law of Hydrocarbons, superseded regional and municipal authorities. The chapter takes a chronological approach to describe how Chávez's discourse informed the creation and implementation of new spatial strategies outlined in the new policy instruments created between 2005 and 2010, as a means of devising new spatial policies to dismantle the existing institutional apparatus of urban governance through the creation of new legal instruments for the construction of a Socialist State Space.

The New Geometry of Power and the construction of the Petro-Socialist State Space

The state-owned oil industry and the country's oil-rich subsoil were a fundamental component of the discursive construction and legitimisation of Hugo Chávez's ideological and political project. He defined the day of his presidential re-election speech of 5 December 2006 as the 'ignition' of a new socialist era:

> The new epoch that begins today will have as its central idea and force the deepening, enlargement and expansion of the Bolivarian revolution, of revolutionary democracy, in Venezuela's life towards socialism.
>
> (BBC Mundo, 2006, translation by the author)

In January 2007 Chávez resumed his weekly Sunday television show *Aló Presidente;* he emphasised that his project was unique because he was building a socialist model different to the socialism that Marx had imagined for European societies, Venezuela's oil wealth and vast oil reserves would be the bulwark of his socialist model (Chávez, 2007a), in other words, he was building his own Petro-Socialism. He declared that the expansion of the Bolivarian Revolution towards Socialism was the only alternative for transcending the evils of capitalism (Chávez, 2007d, p. 63). This announcement was accompanied by the launch of the Orinoco Socialist Project, a roll back of *Apertura Petrolera* (Oil Opening), PDVSA's programme of private investment implemented in the 1990s, thus bringing under the umbrella of the state all operations over the Orinoco Oil Belt's vast extra-crude oil reserves. Bernard Mommer, the Vice-Minister of Hydrocarbons affirmed that oil was a blessing for socialism because it provided the resources that allowed an easy and swift advancement of the transition towards the Socialist State (Chávez, 2007b). Within this rhetoric, the transition towards socialism sustained by oil is presented by Chávez as analogous to a powerful Petro-Socialist vessel fuelled by five revolutionary combustion engines: The Five Engines of the Bolivarian Revolution.

The Petro-Socialist transition is outlined in the National Project Simón Bolívar First Socialist Plan 2007–2013 (PPS), the Five Engines of the Bolivarian Revolution (Chávez, 2007e) ignited in the following sequence:

First Engine: *Ley Habilitante* (Enabling Law, to give special powers to the President to legislate by decree).

Second Engine: *Reforma Constitucional* (Reform of the 1999 Constitution to introduce the socialist rule of law, rejected by popular referendum in 2007).

Third Engine: *Educación 'Moral y Luces'* (Morals and Enlightenment Education, a national public education with socialist values).

Fourth Engine: *Nueva Geometría del Poder* (New Geometry of Power, socialist reorganisation of national geopolitics through the re-distribution of political, economic, social and military power throughout the national territory).

Fifth Engine: *Explosión del Poder Comunal* (Explosion of Communal Power, the implementation and consolidation of the communal state and socialist democracy).

The First Engine, the Enabling Law, is 'the law of laws' that granted Chávez absolute legislative powers to reform laws and create new ones. In other words, it is the law that concentrated all the Bureaucratic Power of the state in his persona. The First Engine ran in sync with the Second Engine, the reform of the

Constitution of the Bolivarian Republic of Venezuela of 1999 (CBRV) to create the model of the new socialist state, expand control over the oil and gas industry while constructing the socialist reorganisation of the political-administrative division of the national territory (Chávez, 2007d, p. 65), consolidating Chávez as the embodiment of the New Magical State. The Third Engine is the transformation of national education according to socialist values imbued with the ideals of Simón Bolívar, Simón Rodríguez and Ezequiel Zamora. The Fourth Engine is the creation of the New Geometry of Power (associated with the Second Engine) to redistribute political, social, economic and military powers over the geographical space of the country according to the new socialist order (p. 67). Finally, the first four engines create the conditions for the final ignition of the Fifth Engine: Communal Power. The climax of this sequence of ignition is the establishment of the Socialist State, and with it the complete dismantlement of the pre-existing institutional apparatus of the Fourth Republic (p. 72).

The Fourth Engine's New Geometry of Power is of particular interest; Chávez affirmed that all these ideas occurred to him while flying over the countryside in a helicopter using binoculars to look below and wondering what could be done to develop the large expanses of uninhabited land which he saw as void of Bureaucratic Power: 'we have great uninhabited spaces where there is no state, where there is no law, and therefore there is no Republic' (Chávez, 2007d, p. 70). The focus of the New Geometry of Power is to eliminate gaps of State Space authority by way of the redistribution of power across the national territory; it involves the creation of a socialist model of federal territories to transition towards the socialist city, a city where communal power rendered obsolete parishes, councils, municipalities and its respective authorities (p. 69). The term is an appropriation of Doreen Massey's 'power-geometry of time-space compression', which concerns power differentials regarding who does and who doesn't control flows and mobility which reflects the unequal distribution of power in relation to time-space compression (Massey, 1991, p. 149, 1994, p. 150). The term time-space compression, first proposed by David Harvey to refer to the acceleration of technology, communications and economic activities that lead to the annihilation of spatial boundaries and distances is used by Massey to speak of 'the geographical stretching of social relations, and to our experience of the world' unevenly distributed under a predominantly westernised and colonised view (Massey, 1994, p. 147). Massey's rhetoric was influential, illustrated by the following extract from Chávez's broadcasted speech during the swearing-in of his new ministerial cabinet on 8 January 2007:

PRESIDENT CHÁVEZ: The geometry, you know it has three dimensions, no? The dimension in line, the distance; the dimension in extension of a territory and the volumetric dimension, the content, the volume. In those three dimensions, I want us to redesign the geometry of power in Venezuela, this will take us to depths, because for example, Apure state, to give you an example, how is Apure state organised? This will direct us to revise organic laws such as that of Municipal Councils, that remain intact and I would say even, it's worse than before.

ATTENDANTS: [applause]

PRESIDENT CHÁVEZ: Municipal Councils that have no power, that are the same old structures, it's the same old fourth republican State.

The regions of the country, how to achieve a symmetrical relationship? Or a symmetrical application of political power, economic power, social power, military power through the length and width of the whole territory. We have many debts there, regions that are too remote, too forgotten, diminished, backward; we have to raise the whole country because it is but one national body. That is why I said that this is an issue we need to develop in depth: the new geometry of power.

(Chávez, 2007c, translation by the author)

Chávez acknowledges the existing deficiencies of Bureaucratic Power over the territory which could potentially endanger the consolidation of the Socialist State, a crucial matter for the longevity of his political project. Lefebvre's notion of State Space is useful to understand this, since territory is a political form of space shaped by historically specific forms of economic and political interventions of the state (Brenner and Elden, 2009), therefore a new socialist order calls for a new territorial order. A new Socialist State could not be consolidated unless it manifested as a Socialist State Space, crucial not only in terms of the Bureaucratic Power of the state but most importantly for securing control and authority over the subsoil from which oil wealth and the magical powers of the New Magical State originate. Chávez's ultimate purpose was to bring into being a new socialist society; he set off to completely restructure the apparatus of the state and in the process demolish the existing bureaucratic structures that were deemed an obstacle or unnecessary.

The National Project Simón Bolívar First Socialist Plan 2007–2013 (PPS), along with the Five Engines of the Bolivarian Revolution, discussed earlier, (Chávez, 2008) is further subdivided into six 'directives':

    I New Socialist Ethic.
   II Supreme Social Happiness.
  III Protagonist and Revolutionary Democracy.
  IV Socialist Productive Model.
   V New National Geopolitics.
  VI New International Geopolitics (Venezuela: New Energy World Power).

The PPS required the replacement of the existing institutional apparatus with the apparatus of the Socialist State. Of special interest for this chapter are directives IV, V and VI, for what they reveal about how the PPS conceptualised the relationship between territory, Petro-Socialism and the consolidation of the Socialist State.

Section IV, the Socialist Productive Model, establishes that the socialist model is founded on the state's total control over productive activities considered strategic for the country's development, the oil industry in particular (Chávez, 2008, p. 43). In this regard, section V, the New National Geopolitics, diagnosed Venezuela's existing 'socio-territorial model' as the by-product of dependency on oil exports managed through a formerly neoliberal PDVSA (Chávez, 2008, p. 57). This 'socio-territorial model' materialised as cities concentrated along the central-northern coastal areas, deemed characteristic of an export-based economy which created a 'structural disarticulation' anchored in a deficient regional and national integration that prioritised connections between ports and centres of primary extraction and the main cities that consume the oil rent; cities that also concentrate large misery belts as well as unregulated human settlements that lack basic services (Chávez, 2008, p. 59). In other words, the PPS speaks of a system of oil cities in Rodolfo Quintero's terms, engendered by a national oil industry dominated by foreign corporations, and more recently, by a neoliberal PDVSA. To overcome these hindrances, the country's 'territorial structure' had to be modified (Chávez, 2008, p. 61). The guidelines of the PPS for the years 2007–2013 laid the foundations for the transition towards a Socialist State Space congruent with Petro-Socialism.

Nonetheless, directive VI, New International Geopolitics, acknowledges that the socialist project is heavily dependent on the oil rent. While directive V denounced the pervasive influence of global oil capitalism in the country's distorted 'socio-territorial model', directive VI affirms that oil will continue to have a significant influence on Venezuela's future, a country that could potentially become an energy world power due to its vast reserves of crude oil. Moreover, it claims that in the face of a world hungry for fossil fuel, Venezuela should not refuse to produce it. Yet again this reveals Hugo Chávez's unrealistic expectations for believing in an 'irreversible trend in the increase of oil prices' (Chávez, 2008, p. 77). In this context, it was the duty of PDVSA to maximise production and increase oil revenue to the state's coffers since the development and longevity of Petro-Socialism and the Socialist State were completely dependent on the global oil market and PDVSA's production capacity. It is evident that Chávez's unrealistic reliance on an inexhaustible supply of oil with sustained high prices informed his discourse and policies. This also reveals an ambition to completely integrate a Socialist Venezuela into global oil capitalism in order to maximise oil revenues, which is in contradiction with his ambition to establish Petro-Socialism as an alternative to capitalism. An essential contradiction runs through Petro-Socialism: Chávez's goal to establish an alternative model to capitalism was heavily reliant on the success of the very model he proposed to eradicate.

However, it becomes coherent in terms of Chávez's exercise of Bureaucratic Power given that the New Magical State is by definition an oil rentier state, its powers and survival reliant on the strength of a national oil industry reframed as revolutionary that provides the resources to consolidate the state's ownership and control over the territory, its subsoil and its natural resources. Chávez's discourse also aimed to reframe the relationship between the state and its territory, but contrary to Lefebvre's conceptual point of the 'territory effect' (Brenner and

Elden, 2009, p. 373) there was no attempt to mask or naturalise the spatial interventions of the state. Chávez was explicit in his ambition to dismantle existing political-administrative structures. His unexpected death in 2013 prompted the publication in 2014 of two posthumous compendiums of his ideas on the New Geometry of Power (*Estado Comunal: La Nueva Geometría del Poder*) and socialist oil policy (*Visión Petrolera de Hugo Chávez Frías, Teoría Socialista sobre la Política Petrolera Venezolana*). The book *Estado Comunal: La Nueva Geometría del Poder* (Communal State: The New Geometry of Power), published by the National Assembly is penned by Manuel Briceño Méndez (2014), professor of Universidad de Los Andes with a PhD in Geography and Ordinance of Territory, at the time representative of the United Socialist Party of Venezuela (PSUV) to the National Assembly and a member of the Permanent Commission of Environment, Natural Resources and Climate Change. Briceño Méndez examines Chávez's political thought retrospectively. He presents the book as a 'humble pedagogical contribution' to the revolutionary debate on the creation of the new Socialist State, an alternative to the destructive model of capitalism, to enable the necessary territorialisation of public policies to achieve 'social collective wellbeing and supreme happiness' (Briceño Méndez, 2014, p. 10). More broadly, he outlines that the cornerstone of Chávez's ideas for the creation of a new socialist society is the territory (p. 11), the creation of a Socialist State Space. Briceño Méndez clarifies in the introduction that his book does not focus on geography but on Chávez's most important contribution: binding to the territory his ideas and theories (p. 11), setting the political framework for the New Geometry of Power as one of the key strategies of the transition towards a new socialist society:

> Socialism has always been interpreted as the political system where the State, the institution that legitimately represents society, controls the means of production, and regulates the participation of citizens in productive processes, particularly the distribution of benefits under principles of equality and solidarity.
>
> (Briceño Méndez, 2014, p. 35, translation by the author)

Therefore, Chávez's Petro-Socialism focuses on the utilisation of the country's natural resources guided by 'socio-cultural rationalities' where production and consumption are tied to the 'principle of felt necessity': instead of satisfying individual 'artificial needs created by improper cultural patterns that respond only to economic profitability and unequal and perverse commercial exchanges', Chávez's intent was for Venezuela to change into 'a society of shared wellbeing underpinned by the socio-cultural principles that identify the Nation' (Briceño Méndez, 2014, pp. 35–37). This transformation would produce a New Geometry of Power to 'consolidate social stability' and establish a 'participatory and protagonist democracy' through the development of a new institutional juridical framework, particularly the reform of 'political-administrative territorial units' and 'public policy management' (Briceño Méndez, 2007, p. 1, 2014, p. 40). His definition of the new institutional juridical framework is modelled after Chávez's seven directives of the PPS (Chávez, 2007e, 2008; Briceño Méndez, 2014, p. 24).

The construction of the New Geometry of Power contemplates the reform of territorial planning to foster a popular power institutionalised in the communal councils and socialist communes (Briceño Méndez, 2014, p. 25). This process would translate in the consolidation of a 'socialist geographical space' with political, social, economic and historical dimensions (pp. 13, 16, 26), evoking the understanding of territory as the 'consequence of specific historical forms of economic and political interventions of the state, which have direct impacts on social life' (Brenner and Elden, 2009, p. 371). Of particular interest is the social agent of this 'socialist geographical space', tied to the utilisation of natural resources and the use of land to satisfy social functions and collective wellbeing. This requires novel modes of direct popular governance and public policy being organised around communal 'territorial units' of population settlement in places with a sense of belonging and 'geographical identity' as an expression of the socio-economic model legitimised by the Socialist State (Briceño Méndez, 2014, pp. 42–43).

The New Geometry of Power involves a dual process of disaggregation of Bureaucratic Power and communal aggregation of popular power throughout the national territory: the communal councils and popular power organisations are grouped in communes; the aggregation of several communes compose the communal city; the communal federation then groups communal cities within a development district; territorial axes and development districts are defined by the State encompassing a number of states and municipalities (Briceño Méndez, 2007, p. 1, 2014, p. 25; López, 2013). Chávez's Second Socialist Plan for the Nation 2013–2019 (Maduro Moros, 2013), pushes further the territorialisation of socialist politics to expand them 'like a giant spider web, to cover the whole territory' (Briceño Méndez, 2014, p. 13). Although Briceño Méndez makes no explicit mention of the national oil industry, referring generally to 'natural resources', it is implicit in Chávez's seventh strategic line: Venezuela a 'world oil power', entrenching the Petro-Socialist model (Chávez, 2007e, 2008). The Petro-Socialist project hinged on the illusion that oil prices would remain high for decades to come.

On the contrary, by 2010 global oil prices were already showing signs of steep decline, which compounded by PDVSA's severe deficiencies caused by the overextension of its duties, mismanagement and lack of re-investment, have translated into a severe decline in oil exports and the glut of oil wealth that Petro-Socialism heavily depended on. Chávez's unrealistic expectations of an enduring oil boom become explicit in the book *Visión Petrolera de Hugo Chávez Frías. Teoría Socialista sobre la Política Petrolera Venezolana* (Oil Vision of Hugo Chávez. Socialist Theory of Venezuelan Oil Policy), authored by economist Andrés Giussepe Ávalo, former deputy for the Andean Parliament, specialist in politics and economics of international commerce of oil, and a regular columnist for revolutionary digital news agency Aporrea.org. The prologue is written by Fernando Soto Rojas, the then President of the National Assembly's Permanent Commission of Oil and Energy, a former guerrilla member of the Fuerzas Armadas de Liberación Nacional (FALN) (Armed Forces of National Liberation) in the 1960s and currently a deputy of the National Constituent Assembly. Giussepe Ávalo's ambition is to formulate Chávez's *cosmovisión petrolera* (oil worldview) through an 'analysis of his speeches, concrete actions, anecdotes, news, strategies, government policies, critiques, proposals, and highly publicised narrated experiences' (Giussepe Ávalo, 2014, p. 21). Giussepe Ávalo claims to find evidence of Hugo Chávez's socialist vision of oil by 'reading between the lines and interpreting his actions' and developing a comprehension that, in his words, provides the theoretical foundation that can be another component of Chávez's political-ideological thought and 'oil worldview' rooted in 'chavismo as a philosophy and a way of taking action in the concrete reality of 21st century Venezuela' (Giussepe Ávalo, 2014, p. 102).

Chávez did not learn about the business of the oil industry until he became president, but in a short period he managed to transform himself into a 'strategist of the geopolitics and the integration in the energetic realm, recognised by leaders, experts and intellectuals worldwide' (Giussepe Ávalo, 2014, p. 37). Regardless of the veracity of this claim, Chávez radically changed oil policy in Venezuela during one of the largest oil rent windfalls in the history of the country. Oil rentierism was necessary to guarantee the irreversibility of Socialism; hence the need for a socialisation of oil capital to use all available financial and economic resources for the promotion and development of social investments (pp. 100–101). Oil was crucial to the advancement of his political project; he insisted that all Venezuelans should know everything related to the key industry that had shaped the country for the past 100 years (Chávez, 2005; Giussepe Ávalo, 2014, p. 37). A century that, according to Chávez, was characterised by the plunder of the country's oil fields, first by foreign corporations and later by a 'nationalised' oil industry with a 'transnationalised vision' (Giussepe Ávalo, 2014, p. 39), casting doubt over Carlos Andrés Pérez's nationalisation of the oil industry in 1976, and a 'transculturised population, marked by the consumerism of the rich and a non-native lifestyle, based on visits and shopping at stores and services similar to those of the society of the United States' (p. 26), echoing Rodolfo Quintero's 'culture of oil' (Quintero, 2011, pp. 19–20) defined as a foreign force of conquest with its own forms of culture and technology that impose a way of life characterised by the exploitation of national oil wealth and the disintegration of local and indigenous culture.

While on the one hand Giussepe Ávalo argues that oil should not be used to satisfy Venezuela's rentier capitalist parasitic economy (Giussepe Ávalo, 2014, p. 28) on the other he concedes that the main objective of Chávez's nationalist oil policy strategies is to secure high oil prices and strengthen rentierism as an economic model focused on the increase in oil rents for the development of other industrial sectors (p. 51), not unlike what previous governments had sought guided by the long enduring slogan of 'to sow the oil'. Chávez adopted the slogan of 'to sow the oil' as it fitted his notion of a socialist oil nationalism. In this vein, oil continues to be first and foremost a matter of sovereignty, a national resource to defend and protect from imperial and foreign powers; to 'rescue' the oil rent and 'revert a history of plunder and bad businesses for the country'; this was a matter of honour for Chávez (p. 76). The 'rescue' of the rent meant that a larger share went to the State coffers, and by extension, to the poorest sectors of society. This is how Chávez consolidated the first phase of his Petro-Socialist revolution: the oil rent trickled down in a bigger proportion to historically neglected sectors of the population (pp. 75–77) by shifting the relationship between PDVSA and the state according to his socialist vision of the oil rent (pp. 102–103). The role of the landlord state persists but with the emergence of 'new oil patriots' (pp. 37–38), oil rentierism is reframed as a socialist oil nationalism in which oil is of 'strategic importance to push a model of political, economic and social development with socialist tendencies'; the rent derived from international oil capitalism is invested to bolster the revolutionary process of creating a new socialist society (pp. 24, 103). In the remit of the national oil industry, the Executive's exercise of national oil policy (Hugo Chávez, and currently Nicolás Maduro Moros) is mediated by the Ministry of Popular Power for Energy and Oil, and of all operational and financial activities administered by PDVSA reimagined as a revolutionary enterprise subordinate to the Socialist State.

The CBRV, the development plans for the nation, and the new laws created in the wake of a socialist oil nationalism obeyed Chávez's will to create a new Socialist State Space based on the New Geometry of Power. Oil provided the resources to push forward the creation of Federal Provinces, Functional Districts, Special Military Regions and Special Authorities to displace the authority of municipalities, parishes and mayoralties; a rollback of the politico-territorial decentralisation process of 1989–1999 (Delfino, 2005, pp. 22–31) characterised by Chávez as an imperialist strategy to split up the territory in order to weaken and decapitate the state:

> All a great strategy for weakening national power, national unity. The revolutionary process must go in the opposite direction, it has to strengthen national unity. I have never liked the word decentralisation, you know why?, because it sounds like decapitation, decapitate, take the head off, decentre, removing the centre, and everything needs a centre.
>
> (Chávez and Harnecker, 2005, p. 69, translation by the author)

The process of transformation of the legal framework of territorial organisation to displace the apparatus under which regions, states, cities, municipalities and parishes were governed under the premises of decentralisation was mired with discrepancies and contradictions. To unpack the web of strategies deployed by the New Magical State to dismantle and reform the institutional apparatus while creating new legal entities that conceptualise the Socialist State Space, a detailed chronological account is needed to trace the myriad of institutional changes brought by Chávez as he disassembled the Bureaucratic Power of the 'counter revolutionary state' at all scales, from the Constitution to municipal laws. To remain within the scope of this book, the chronological account will focus primarily on Caracas, as it is the physical seat of Bureaucratic Power of the state and in particular, where most of PDVSA La Estancia's interventions concentrate.

PDVSA's Oil Social District as a parallel Petro-Socialist State Space

The fast pace of abrogation and creation of new laws produced institutional instability while the lack of continuity augmented existing bureaucratic deficiencies. The case of Caracas, whose fragmented legal and institutional framework already carried deficiencies inherited from previous governments serves as an exemplar to illustrate the discontinuities of the transition and the institutional mechanisms that ultimately enabled PDVSA to instrumentalise the Organic Law of Hydrocarbons to supersede the legal authority of municipalities and in the process, interpret and implement it as an implicit cultural policy.

The CBRV was the first constitution approved by popular referendum in Venezuelan history. It displaced the Constitution of 1961 and inaugurated the era defined by Chávez as the Fifth Republic. Among the innovations of the CBRV were the change of the name of the country from the Republic of Venezuela to the Bolivarian Republic of Venezuela and the reform of the structure of government that established a decentralised government to grant greater powers to the legislative branch (King, 2013, p. 379). The CBRV also reformed the institutional structure of the capital city. The juridical framework of the territory of Caracas has been fraught with weaknesses and dispersion in terms of policies that define the material space of the city, its administrative boundaries and bureaucratic structures (Delfino, 2001, pp. 36–40; Negrón, 2001, p. 11) evidenced by severe deficiencies in urban management, as each municipality works autonomously. Furthermore, the reforms in the Constitution and subsequent decisions of Chávez's government did little to remedy these weaknesses; on the contrary, they were magnified as the institutional and juridical dispersion was fertile ground to implement with ease the New Magical State's territorial strategies.

Article 18 of the CBRV ratified Caracas as the seat of national power and specified that a special law had to be created to 'establish the territorial and political unity of Caracas' (Asamblea Nacional Constituyente, 1999). Article 16 of the CBRV substituted the Federal District with the Capital District, mainly circumscribed to the boundaries of Libertador Municipality. The Capital District is the only district in the country under a special regime, it is a special political-territorial entity that constitutes the capital of the republic but has attributions similar to the other 23 states (Gobierno del Distrito Capital, no date). The Capital District is the permanent seat of the Presidency, many ministries, the National Assembly, the Supreme Court, the Comptroller General, as well as the Joint Command of the Armed Forces and the Marine, Air Forces and the Bolivarian National Guard.

Although the CBRV only mentions the creation of the Capital District, Article 18 was interpreted as the foundation to create the Metropolitan District of Caracas, put in effect in 2000 through the Special Law for the Regime of the Metropolitan District of Caracas (LERDMC), legislated by the National Assembly (Delfino, 2002, p. 134). The LERDMC established that the material space of the Metropolitan District is composed of Libertador Municipality/ Capital District and the municipalities of Chacao, Sucre, Baruta and El Hatillo (Asamblea Nacional de La República Bolivariana de Venezuela, 2000). The Metropolitan District was defined as an entity that guarantees the territorial unity of Caracas, which is confusing because the law simultaneously establishes the preservation of the territorial integrity of Miranda State, which also has sway over El Hatillo and Baruta (Delfino, 2002, p. 137). The Metropolitan District was given the same competencies as municipalities, overlapping attributions and competencies with Miranda State, rendering the municipalities redundant. More-over, the law was perceived by municipal authorities as a threat to their auto-nomy (pp. 135, 142). As a result of the LERDMC, four different spheres of government coexist and clash within Caracas: national government, Miranda State, Metropolitan District and Capital District (Delfino, 2001, p. 40).

In 2007 Hugo Chávez put forward a referendum for the reform of the CBRV. The reform was conceived as an instrument for the dismantlement of the 'constitutional and legal superstructure' that had sustained the capitalist mode of production, in order to embark on the construction of a socialist society for the twenty-first century and the establishment of a New Geometry of Power. Although he lost the referendum vote, all the laws, decrees and policy instruments created before and after Chávez's re-election in December 2006 implemented the repealed constitutional reform by stealth. Even more, the legal foundations for the Socialist State had already been laid out by the National Assembly in clear breach of the CBRV, with the sanction in 2006 of the *Ley de Consejos Comunales* (Law of Communal Councils), reformed and elevated to the status of Organic Law in 2009 (Brewer Carías, 2011, p. 127). In December 2010, a month before the newly elected National Assembly took power with a larger representation of the opposition, a number of organic laws were swiftly sanctioned (p. 128) to establish the legal framework of the Social-ist State: Organic Law of the Popular Power of the Communes, Organic Law of the Communal Economic System, Organic Law of Public and Communal Planning, as well as reforms to the Organic Law of Municipal Public Power and the Organic Law of Local Councils of Public Planning. These laws reorganised the bureaucratic structure of the state under the principles of Socialism to establish a communal economic system that contradicts, and runs in parallel to, the mixed economic structure and economic liberties established by the CBRV. The Commune, defined as an entity of direct popular sover-eignty, is the backbone of the socialist order, meant to displace the muni-cipality as the primary entity of territorial organisation (p. 129). Under this principle, municipalities are meant to surrender their Bureaucratic Power and authority to Communal Councils (p. 129).

The deficiencies in coherent spatial policies for Caracas were compounded with the transition towards the Socialist State Space, evidenced by the journey from bill to legal vacuum of the Organic Law for the Planning and Management of the Organisation of Territory (LOPGOT). The LOPGOT created in 2005 abrogated the Organic Law for the Organisation of Territory of 1983 and the Organic Law of Urban Planning of 1987, the key spatial policy instruments that shaped the role of institutions with competencies in urban planning and management in Caracas, and the country at large. The publication of the LOPGOT in the Official Gazette of the Bolivarian Republic of Venezuela included a *vacatio legis* of six months to come into effect in March 2006. In March 2006, a partial amendment to the law was published with another *vacatio legis* of six months that deferred its enactment to September 2006. The day the amendment was due to come into effect, a second amendment was published with another *vacatio legis* of six months with a specific date: 28 February 2007 (Brewer Carías, 2007, p. 2). In February 2007 it was repealed by the Organic Law for the Repeal of the Organic Law for the Planning and Management of Territory which consisted of only two articles. Article 1 repeals the law while Article 2 states that the repeal would come into effect from the date of its publication in the Official Gazette. The announced publication in the Official Gazette never happened. The twice deferred LOPGOT of 2005 and the two partial amendments meant to substitute the law of 1983 never came into effect and remained in a suspended *vacatio legis* that left a vacuum in the regulation of territory and urban planning. It is in the breach left between the suspended *vacatio legis* that lingered from 2005 and the creation of the legal framework for the Communal State in 2010 that the interventions in the city by PDVSA La Estancia take place. This fragmented and dispersed institutional landscape is the context of the extension of PDVSA's State Space, the Oil Social District, over the Metropolitan District, which allowed PDVSA La Estancia to intervene in the restoration of public art and public spaces across the city, displacing the authority of municipalities. PDVSA La Estancia managed to override the unstable institutional dispersion described previously by abiding to the Organic Law of Hydrocarbons, Article 5 and the Oil Social Districts.

The Oil Social District derives from the 'old' PDVSA's Oil Districts, operational units of territorial management of areas of oil extraction, oil camps and oil fields. After the oil strike of 2002–2003, the Oil Districts were renamed Oil Social District, to delineate the social mission of the revolutionary new PDVSA, aligned with the principles of the PPS. By definition, the Oil District should only cover areas of extraction, production and refinement of hydrocarbons but with the Oil Social District the definition is expanded to a 'unit of territorial management with productive, ecological and social character that within the process of exploration, production, refining and commercialisation of oil and gas' guarantees endogenous sustainable development to 'generate the balances needed to eradicate poverty' and wider access to education and health care (Ramírez, 2005, p. 2; Figueredo, 2006). In sum, it establishes that communities surrounding oil fields must benefit from social programs and infrastructures funded and built by

PDVSA. In total, 17 Oil Social Districts cover 20 of 23 states plus the Capital District (Rodríguez, 2006):

1. Maracaibo District.
2. Tia Juana District.
3. Lagunillas District.
4. Tomoporo District.
5. Barinas District.
6. Apure District.
7. Paraguaná District.
8. El Palito District.
9. Metropolitano (Metropolitan) District.
10. Puerto La Cruz District.
11. Delta Caribe Oriental District.
12. Anaco District.
13. Norte (North) District.
14. San Tomé District.
15. Morichal District.
16. Delta District.
17. Faja (Oil Belt) District.

Oil Social Districts cover most of the national territory. The Metropolitan Oil Social District extends from central Guárico State in the south to the north to cover Aragua State, Miranda State and the Metropolitan District.

The subsoil of the Caracas valley does not contain crude oil reserves, nor has the city ever been a site of crude oil extraction or refinement. Nonetheless, the Metropolitan Oil Social District encloses Caracas because it is where the corporate headquarters of PDVSA are located. Thus, the headquarters of PDVSA are conceptualised as a centre of oil extraction and distribution of oil rent right in the midst of Libertador Municipality, effectively reframing Caracas as an oil field. The legal vacuum left by the transition towards the Socialist State Space also created the conditions to conceptualise and enact the Oil Social District as PDVSA's parallel State Space. This is what enabled PDVSA La Estancia to absorb Caracas into the sphere of influence of PDVSA and undertake its works of public art and public space restoration across the city.

The process of dismantlement and disaggregation of Bureaucratic Power prescribed by the PPS was concerned with the abrogation of the existing legal instruments of territorial management and the creation of the legal entities of the Socialist State Space. The process was far from coherent, fraught with inconsistencies that further fragmented and eroded the already diminished state capacity in urban governance, particularly in Caracas. Although it did not translate into an actual physical radical transformation of Caracas, it had clear and palpable implications in generating an overlap of conflicting policies and the ensuing perception of institutional chaos. This is important as the overlaps and discrepancies between the abrogated policy instruments, PDVSA's de facto State Space and the Socialist State were instrumental in enabling PDVSA La Estancia to override the authority of municipalities by abiding to the Organic Law of Hydrocarbons. This allowed PDVSA La Estancia to appropriate the material space of Caracas (and other cities in Venezuela) re-conceptualised as an oil field enclosed by the Oil Social Districts, made manifest physically in the city through works of restoration, urban regeneration and discursively in the adverts through a notion of Culture as Renewable Oil that binds culture to the territory, culture conceived metaphorically by the state-owned oil company as a material entity that accumulates in the subsoil, under the tight control of the Petro-Socialist state.

Conclusion

The entanglements between Hugo Chávez's discourse and the creation of the new spatial strategies of the Socialist State Space were a mobilisation of his Bureaucratic Power to completely remodel socio-spatial relations according to his political project. The power of Chávez's discourse to translate into action the reform of the institutional apparatus of the state originated on his embodiment of a New Magical State that concentrated all the Bureaucratic Powers of the state in his persona. Furthermore, his model of Petro-Socialism was informed by unrealistic expectations of an inexhaustible supply of oil and high revenues that emphasised the contradiction of constructing a Socialist State heavily dependent on the success of global oil capitalism. Chávez did not envision a post-oil future for Venezuela. Thus, the survival of the Socialist State became inextricably dependent on a direct life line to global oil capitalism, the system he set out to challenge within Venezuela's borders when he outlined his political platform as an alternative to neoliberalism. Chávez's reliance on oil rentierism for the consolidation of the Socialist State reveals an essential contradiction that underpinned Petro-Socialism: its reliance on the success of the economic model it was meant to eradicate. Nonetheless, the contradiction gains coherence in terms of Chávez's 'dramaturgical' exercise of Bureaucratic Power, given that only an oil rentier state can be a Magical State and only oil can provide the vast resources needed to embark on a total reform of the nation's institutional apparatus to implement the new spatial strategies of the Socialist State and consolidate absolute ownership and control over the territory, its subsoil and its natural resources. In this sense, territory was essential to Petro-Socialism. The chronological approach used to unpack the process of abrogation, creation and amendments of the legal instruments of territorial organisation guided by the premises of the New Geometry of Power, served to illustrate a process fraught with inconsistencies that entrenched diminished state capacity and created a chaotic bureaucratic landscape in urban governance, particularly in the case of Caracas.

The New Geometry of Power conceptualised the political and administrative boundaries of the Socialist State Space, which came into conflict with the existing institutional apparatus and the CBRV, creating fault lines within the bureaucratic structure that enabled PDVSA La Estancia to interpret and implement the Organic Law of Hydrocarbons as both a territorial policy and implicit cultural policy instrument. The New Geometry of Power remains an incomplete project (López, 2017), notwithstanding the profound changes Chávez made to the juridical framework of territory ordinance.

The new spatial strategies were devised as an abrupt break, a disruption of the institutional order inherited from previous governments, to undermine and eventually completely substitute the existing apparatus. The vacuum left by the delay in pushing forward the new legal instruments of the Socialist State, generated a bureaucratic landscape in which the only remaining stable legal framework and institution was PDVSA's. It is PDVSA, as a de facto parallel State Space, that succeeds in the territory effect, not just by naturalising its sway over the city by instrumentalising the Oil Social District to carry out works of restoration of public art and urban regeneration, but also by concealing it through the visual construction of the adverts, analysed in Chapter 5, in which giant oil workers become a naturalised sign of PDVSA's State Space.

Chapter 4 expands this discussion by looking in particular at the speeches of Hugo Chávez, the former President of PDVSA Rafael Ramírez and the former General Manager of PDVSA La Estancia, Beatrice Sansó de Ramírez. It explores the discursive strands that run through the state-owned oil company's narrative of its commitment with Petro-Socialism, assuming the 'magical' power of marvel of the New Magical State confined to the realm of oil around the narratives of 'sowing oil' and culture as 'renewable oil'. It also examines how PDVSA La Estancia's urban management of city space compensates for the state's diminished capacity, and how through its direct access to the oil rent, it was able to extend its own dominant space over Caracas and the country at large, aided by the interpretation of Article 5 of the Organic Law of Hydrocarbons as an implicit cultural policy.

References

Asamblea Nacional Constituyente (1999) 'Constitución de la República Bolivariana de Venezuela'. Caracas, Venezuela: Gaceta Oficial de la República Bolivariana de Venezuela No. 36.860.

Asamblea Nacional de La República Bolivariana de Venezuela (2000) 'Ley Especial sobre el Régimen del Distrito Metropolitano de Caracas'. Caracas, Venezuela: Gaceta Oficial de la República Bolivariana de Venezuela No. 36.906.

BBC Mundo (2006) 'Chávez presidente reelecto', *BBC Mundo*, 5 December. Available at: http://news.bbc.co.uk/hi/spanish/latin_america/newsid_6205000/6205158.stm.

Brenner, N. and Elden, S. (2009) 'Henri Lefebvre on State, Space, Territory', *International Political Sociology*, 3(4), pp. 353–377 (accessed: 28 August 2015).

Brewer Carías, A. (2007) 'Las limitaciones administrativas a la propiedad por razones de ordenación territorial y ordenación urbanística en Venezuela, y el curioso caso de una ley sancionada que nunca entró en vigencia', *Revista do Direito*. Curitiba, Brazil: II Congreso Ibero-Americano de Directo Administrativo, 42(Jan–Abr 2014), pp. 120–153.

Brewer Carías, A. (2011) 'Las Leyes Del Poder Popular Dictadas En Venezuela En Diciembre De 2010, Para Transformar El Estado Democrático Y Social De Derecho En Un Estado Comunal Socialista, Sin Reformar La Constitución', *Cuadernos Manuel Giménez Abad*, 1, pp. 127–131.

Briceño Méndez, M. (2007) 'La ciudad en la ordenación del territorio y la Nueva Geometría del Poder, I Seminario sobre la Ciudad Sustentable ULA'. Mèrida, Venezuela.

Briceño Méndez, M. (2014) *Estado Comunal: La Nueva Geometría del Poder*. Caracas, Venezuela: Fondo Editorial de la Asamblea Nacional Willian Lara.

Chávez, H. (2005) 'Aló Presidente #220'. Venezuela: Venezolana de Televisión. Available at: www.alopresidente.gob.ve/materia_alo/25/1279/?desc=Alo_Presidente_220. pdf (accessed: 28 August 2015).

Chávez, H. (2007a) 'Aló Presidente #263'. Venezuela: Venezolana de Televisión. Available at: www.alopresidente.gob.ve/galeria/23/p-16/tp-46/ (accessed: 28 August 2015).

Chávez, H. (2007b) *Aló Presidente #288, www.alopresidente.gob.ve*. San Diego de Cabrutica, Venezuela: Sistema Bolivariano de Comunicación e Información SIBCI. Available at: www.alopresidente.gob.ve/materia_alo/25/1396/?desc=nro[1]._288_alo_presidente_-_28-jul-2007__estado_anzoategui___corregido_ljc_.pdf (accessed: 28 August 2015).

Chávez, H. (2007c) *Intervención del Comandante Presidente Hugo Chávez durante acto de juramentación de nuevo Gabinete Ejecutivo*. Caracas, Venezuela. Available at: http://todochavez.gob.ve/todochavez/2715-intervencion-del-comandante-presidente-hugo-chávez-durante-acto-de-juramentacion-de-nuevo-gabinete-ejecutivo (accessed: 9 July 2018).

Chávez, H. (2007d) 'Juramentación del Presidente de la República Bolivariana de Venezuela, Hugo Chávez Frías (período 2007–2013)'. Caracas, Venezuela: Asamblea Nacional de la República Bolivariana de Venezuela.

Chávez, H. (2007e) 'Proyecto Nacional Simón Bolívar Primer Plan Socialista – PPS – Desarrollo Económico y Social De La Nación 2007–2013'. Caracas, Venezuela: Presidencia de la República Bolivariana de Venezuela.

Chávez, H. (2008) 'Líneas Generales del Plan de Desarrollo Económico y Social de la Nación 2007–2013'. Caracas: Ministerio del Poder Popular para la Comunicación y la Información.

Chávez, H. and Harnecker, M. (2005) *Taller de Alto Nivel 'El nuevo mapa estratégico' 12 y 13 de noviembre de 2004. Intervenciones del Presidente de la República Hugo Chávez Frías*. Edited by M. Harnecker. Caracas, Venezuela.

Coronil, F. (1997) *The Magical State: Nature, Money, and Modernity in Venezuela*. Chicago: University of Chicago Press.

Corrales, J. and Penfold, M. (2011) *Dragon in the Tropics. Hugo Chávez and the Political Economy of Revolution in Venezuela*. Washington D.C.: The Brookings Institution.

Delfino, M. de los Á. (2001) 'La gobernabilidad de Caracas Capital y el Distrito Metropolitano', *Urbana*, 6(29), pp. 35–45.

Delfino, M. de los Á. (2002) 'Reflexiones sobre el Distrito Metropolitano de Caracas', *Revista Venezolana de Economía y Ciencia Sociales*, 8(3), pp. 131–149.

Delfino, M. de los Á. (2005) 'El proceso de decentralización político-territorial venezolano y su desarrollo constitucional y legislativo', *Urbana*, 36, pp. 13–33.

Figueredo, C. (2006) *Oportunidades y Retos Costa Afuera, Asociación Venezolana de Procesadores de Gas*. Available at: www.venezuelagas.net/documents/2006-PL-08-spa.pdf (accessed: 11 July 2018).

Giussepe Ávalo, A. R. (2014) *Visión petrolera de Hugo Chávez Frías. Teoría Socialista sobre la Política Petrolera Venezolana*. Caracas, Venezuela: Editorial Metrópolis.

Gobierno del Distrito Capital (no date) *¿Por qué el Distrito Capital?* Available at: www.gdc.gob.ve/content/site/module/pages/op/displaypage/page_id/3/format/html/ (accessed: 2 November 2015).

King, P. (2013) 'Neo-Bolivarian Constitutional Design Comparing the 1999 Venezuelan, 2008 Ecuadorian and 2009 Bolivian Constitutions', in Galligan, D. J. and Versteeg, M. (eds) *Social and Political Foundations of Constitutions*. Cambridge, MA: Cambridge University Press, pp. 366–397.

López, E. (2013) 'La nueva geometría del poder quedó inconclusa', *El Nacional*, 6 March. Available at: http://seminariogp.blogspot.com/2013/07/la-nueva-geometria-del-poder-quedo.html (accessed: 29 June 2018).

López, E. (2017) 'La nueva geometría del poder quedó inconclusa', *El Nacional*, 11 April. Available at: www.el-nacional.com/noticias/politica/nueva-geometria-del-poder-quedo-inconclusa_130456 (accessed: 11 July 2018).

López-Maya, M. (2007) *Nuevo debut del Estado mágico, Aporrea.org*. Caracas, Venezuela. Available at: www.aporrea.org/actualidad/a35326.html (accessed: 8 December 2015).

López-Maya, M. (2013) 'El incierto porvenir del Estado comunal', *SIC*, 753, pp. 101–102.

Maduro Moros, N. (2013) *Plan de la Patria Segundo Plan Socialista de Desarrollo Económico y Social de la Nación, 2013–2019*. Caracas, Venezuela: Asamblea Nacional de la República Bolivariana de Venezuela.

Massey, D. (1991) 'A Global Sense of Place', *Marxism Today*, 38, pp. 24–29.

Massey, D. (1994) 'A Global Sense of Place', in *Space, Place and Gender*. Cambridge: Polity, pp. 146–156.

Negrón, M. (2001) 'Caracas en busca de la gobernabilidad', *Urbana*, 6(29), pp. 9–12.

Quintero, R. (2011) 'La cultura del petróleo: ensayo sobre estilos de vida de grupos sociales de Venezuela', *Revista BCV*, pp. 15–81.

Ramírez, R. (2005) 'Nuestro Compromiso es gerenciar una empresa nacional en tiempos de revolución. Discurso No. 5, 11 de Julio de 2005'. Caracas, Venezuela: Ministerio de Energía y Petróleo.

Rodríguez, F. (2006) 'Distritos Sociales con visión nacional, popular y revolucionaria, Asociación Venezolana de Procesadores de Gas'. Available at: www.venezuelagas.net/documents/2006-PL-10-spa.pdf (accessed: 11 July 2018).

4 Bureaucratic Power, performative speech and oil policy

'Sow the oil' to 'harvest culture'

The oil windfall brought by the oil concessions in the 1930s, and later by the oil boom that followed the nationalisation of the oil industry in the 1970s, prompted Venezuelan novelist, intellectual and politician Arturo Uslar Pietri to urge the nation's leaders and elites 'to sow the oil'. The slogan 'to sow the oil' is at the heart of enduring and conflicting views around oil in Venezuela (Pérez Schael, 1993; Pérez Alfonzo, 2011; Quintero, 2011). Uslar Pietri used farming language as a didactic device to suggest the manner in which oil should be invested, by making reference to the riches of the land rather than to oil as an immaterial and ephemeral source of wealth (Pérez Schael, 1993, pp. 199–205). During the first half of the twentieth century, foreign oil companies established in Venezuela intervened in national politics and reorganised the territory both in rural and urban areas, shaping the attitudes of subsequent generations towards the oil industry – whether national or foreign – as oil became akin to progress, development and the emerging modern nation (Tinker Salas, 2014, p. 18). In the aftermath of the collapse of oil prices in the 1980s and 1990s (the period that led to the rise of Hugo Chávez), sowing oil had become an unfinished utopia.

Nonetheless, Hugo Chávez's adoption of Uslar Pietri's slogan to baptise his national oil policy *Plan Siembra Petrolera* (Sowing Oil Plan) is used as a discursive construction to portray Petro-Socialism as a successful strategy for sowing oil. The Sowing Oil Plan was launched in 2005 as a 25-year strategy, but with the passing of Hugo Chávez in 2013 it seemed that only PDVSA had the means and resources to exert the Bureaucratic Power of the New Magical State. While Petro-Socialism was the vehicle to constitute the Socialist State Space, the disarticulated process to constitute the new legal instruments of the Socialist State Space enabled PDVSA to expand the Oil Social District as a dominant parallel State Space.

If the city of Caracas is the material space of the state's Bureaucratic Power, this chapter argues that PDVSA, as an instrument of Petro-Socialism, exercised its power over Caracas by envisioning the city as a Petro-Socialist urban oil field. Hence, if there is a vision of the city of the future in Petro-Socialism, then it is PDVSA (and PDVSA La Estancia) who possess the 'magical' Bureaucratic Power to enact it.

PDVSA La Estancia takes its name from its main building, the eighteenth century colonial house of Hacienda Estancia La Floresta, a former coffee and sugarcane plantation that was privately urbanised in the 1940s and 1950s to become the upper middle-class neighbourhood of La Floresta. Only the house and its adjacent gardens have survived. By the 1950s and 1960s La Estancia was surrounded by middle and upper middle-class neighbourhoods and modern office buildings such as the headquarters of Mobil and the United States Embassy. The hacienda house and gardens were purchased by PDVSA in the 1980s with the purpose of demolishing it and building on its site a new purpose built high rise office complex to relocate its headquarters (Grauer, 2015). The project for the new headquarters never materialised, and the company had to decide what to do with the hacienda it owned (Grauer, 2015). In 1990 La Estancia was declared a National Historic Monument; PDVSA took on the restoration and refurbishment of the hacienda house and gardens to adapt its use as a private venue for corporate events and accommodation for VIP guests. The Centro de Arte La Estancia opened in 1993, a cultural centre for the arts and industrial design following the advice of experts from Galería de Arte Nacional and the artist-designer Miguel Arroyo (Sato Kotani, 2015). The centre's cultural activities were accompanied by a programme of community outreach aligned with PDVSA's strategy of corporate social responsibility (Grauer, 2015). With the exception of private corporate events, Centro de Arte La Estancia was open to the public free of charge (Sato Kotani, 2015).

After the events of 2002–2003 (the failed coup d'état and national oil strike), Hugo Chávez announced the rise of a 'new PDVSA' ideologically aligned with the Bolivarian revolution. Coinciding with the launch of the Sowing Oil Plan 2005–2030, PDVSA's Centro de Arte La Estancia (La Estancia Art Centre), regarded as an elitist institution by the revolution, reopened its doors in 2005 as PDVSA La Estancia, the social and cultural arm of the new PDVSA.

PDVSA La Estancia described itself as 'an oasis of culture and knowledge' (PDVSA La Estancia, no date a) for 'the appreciation, restoration, promotion and dissemination of the historical and artistic heritage of the country' (PDVSA La Estancia, no date b) guided by three axes of action: social, cultural and heritage. The institution had become a reference within the country for the scope and quality of their work and modern aesthetics of the spaces and architecture they have intervened. These works were able to take place because the institutional and bureaucratic framework of city governance was weakened by a simultaneous process of fragmentation and centralisation of Bureaucratic Power.

The chapter is divided into three parts. First, it develops the critical discourse analysis framework utilised to examine the relationship between power, discourse and performative utterances. Second, this framework is deployed to examine public speeches of the three leading figures of the national oil industry between 2005 and 2014: late President Hugo Chávez, former president of PDVSA Rafael Ramírez and former General Manager of PDVSA La Estancia, Beatrice Sansó de Ramírez. It examines the articulations between Chávez's discursive construction of Petro-Socialism with the public speeches of Rafael Ramírez; followed by the speeches of the General Manager of PDVSA La Estancia, Beatrice Sansó de Ramírez, to flesh out two discursive strands: one, PDVSA La Estancia as an instrument of the Sowing Oil Plan that 'harvests culture' and two, the 'utopia of the possible'. This chapter draws on Zygmunt Bauman (2004) and Jeremy Ahearne (2009) to demonstrate how these constructions, built on the stratum of the disjointed process to constitute the Socialist State Space, enabled PDVSA La Estancia to interpret Article 5 of the Organic Law of Hydrocarbons as an instrument of implicit cultural policy.

The performative power of discourse

This chapter engages with critical discourse analysis (CDA) because it is interested in the way in which language and discourse are used to achieve social goals and the part this use plays in social change, in relation to Venezuela's transition towards socialism and the New Magical State's 'dramaturgical' (Coronil, 1997, p. 239) exercise of Bureaucratic Power through his performative 'discourse-spectacle' (Pinardi, 2007). In CDA the assumption is that ideas and practices that constitute such discourses transmit, produce and reproduce power, but also undermine and expose its fragility. CDA is not an analysis of discourse as a text but the analysis of the 'relations between discourse and other elements of the social process' as well as the internal relations of discourse (Fairclough, 2010, p. 4,10). CDA does include the systematic analysis of texts, but it is not just descriptive, it is also normative; the analysis of these complex relations 'cuts across conventional boundaries between disciplines' making CDA the ideal trans-disciplinary form of analysis for this study. CDA is also relational and dialectical, its primary focus is on discourse as a complex set of social relations, which can be also 'layered' including relations between relations (p. 3) such as the dialectical relations between discourse and power (p. 8).

Discourse cannot be defined as a separate entity nor as a mere expression of social practice, it can only be understood 'by analysing sets of relations', as discourse 'brings into the complex relations which constitute social life: meaning, making meaning' (Fairclough, 2010, p. 3). Discourse can also be defined 'as the flow of knowledge' through history, a flow that 'determines individual and collective doing and/or formative action that shapes society, thus exercising power' (Jager, 2011, p. 34). Discourses can exert power because they are regulated, institutionalised and linked to action; as the world is construed discursively, whether these construes exercise 'socially constructive effects' is dependent on power relations (Fairclough, 2010, pp. 4–5). Although power and discourse can be identified as different moments of the social process they are entangled as 'power is partly discourse, and discourse is partly power' (p. 3); discourses condense and simplify the complexities of power relations.

Power relations bring questions of ideology, understanding ideology as meaning in the service of power and domination (Thompson, 1984, p. 132), forms of representations of the world that 'may be operationalised in ways of acting and interacting' that contribute in the establishment and perpetuation of unequal relations of power (Fairclough, 2010, p. 8). The power of discourse is linked to different mechanisms of power and institutions, as discourses both shape societal reality and are a reflection of reality. Hence, discourses can be defined as material realities, they shape reality through 'intervening active subjects in their societal contexts as (co-)producers and (co-)agents of discourses' (Jager, 2011, p. 36). These active interventions also construe discourses as 'societal means of production':

> discourses exercise power. They are themselves a power factor by being apt to induce behaviour and (other) discourses. Thus, they contribute to the structuring of the power relations in a society.
>
> (Jager, 2011, p. 37)

Thus, the character of Bureaucratic Power, exercised by individuals in control of the state apparatus, is partially discursive. In the context of this chapter, CDA is concerned with the analysis of the production of reality performed by discourse, conveyed by active people in power: Hugo Chávez, Rafael Ramírez and Beatrice Sansó de Ramírez (Jager, 2011, p. 36). In this vein, their discourses are also regarded as a means for the production of the Socialist State, PDVSA's State Space and the notion of Culture as Renewable Oil.

Siegfried Jager (2011, p. 47) proposes eight topics to identify the structure of discourses. *Discourse strands* are flows of discourse that concentrate on a common topic, each discourse strand can also contain a number of subtopics that could be summarised into groups of subtopics. *Discourse fragments* are texts, or parts of a text, that covers a certain theme, discourse fragments can combine to form a discourse strand. *Entanglements of discourse strands* occur when a text can make references into several discourse strands: 'in a text various discourse fragments can be contained, these emerge in general and entangled form' (Jager, 2011, p. 47). The overall societal discourse in its entanglement and complexity take into consideration that societies are not homogeneous, therefore, the several discourse strands that flow from the general societal discourse are 'in a state of complex entanglement' (p. 50). *Discursive events* and *discursive context* refer only to events that can be politically emphasised to be considered discursive events (p. 48). *Discourse planes* refer to the societal locations (media, science, business, everyday life, politics, etc.) from which 'speaking happens' since 'discourse strands operate on various discursive planes' (p. 49); it needs to bear into consideration that all discourse planes are closely entangled. *Discourse position* refers to the specific ideological location of the individual or medium. *History, present and future of discourse strands* establishes that all discourses and discourse strands 'have a history, a present and a future', therefore it is necessary to consider the longer timeframes of discursive processes to determine their 'genealogy'. This sort of endeavour is extremely ambitious as it would entail the undertaking of a large number of strands of single projects, but such a scale falls outside of the scope of this study as it does not encompass the history of the discourse strands that will be identified in the analysis nor the overall entanglements with other societal discourses.

The analysis is confined to the topics of Bureaucratic Power, State Space and Culture as Renewable Oil in the time frame of Chávez second and third term in office (2005–2013). Part of the analysis of the speeches will draw on a selective use of J. L. Austin's speech-act theory, looking in particular at performative utterances. For Austin, all sentences form a class of utterances, each class defined grammatically. He contrasted constative and performative utterances, but he ultimately argued in his 1958 paper 'Performatif-Constatif' that there is no precise distinction between performative and constative utterances reducing all statements to performatives. However, Jacobsen (1971, p. 357) argues that Austin's relativism was produced by basing the distinction on a purely verbal and grammatical criterion. A performative utterance is then:

> inseparably connected with the fact that an act is either performed explicitly by a person, as e.g. 'I promise you', or by a group of persons authorized to act as one person, as e.g. 'You are requested to shut the door', where the request has to be issued either by one person or by a group of persons (officials), who have the authority to make requests concerning the regulations governing the public behavior.
>
> (Jacobsen, 1971, p. 359)

Jacobsen proposes that although it is true that an act is always performed by a person it does not necessarily follow that a person performs an act by what that person says, in saying something it only follows that the person performs the act of speech. A constative utterance is transferable because it refers to more general statements; for example, in the statement 'I am Venezuelan' the 'I' is transferable to any person the description applies to.

On the other hand, a performative utterance is a non-transferable 'speech act which does not merely describe something but enacts it' (Cuddon, 2013, p. 525), it has some degree of inherent agency. The non-transferability of the performative utterances stemming from the speech act of a person or group of persons with the authority to influence or govern public behaviour (officials) is central to the analysis of the public speeches by Hugo Chávez, the President of PDVSA and the General Manager of PDVSA La Estancia to examine how their discourses exert power to enact the shaping of State Space and culture.

Jager proposes an analytical guideline to follow for the critical analysis of discourses. His guideline is designed to cater to the specific problems of media analysis (Jager, 2011, p. 54); the main medium of diffusion or publication of Chávez's, Ramírez's and Sansó de Ramírez speeches was a mix of mass media and digital platforms: television broadcasts, a personal blog of the institution's key authority, the institution's official website, Facebook and YouTube. For the purpose of this analysis, Jager's guideline has been adapted in the following form:

1 General characterisation of the speaker and speeches.
2 Overview of textual material (speeches selected and criteria for selection, identification of thematic areas or discursive strands, as well as possible discourse strand entanglements).

3   Analysis of relevant discourse fragments (institutional framework/context, justification for the selection of fragments, selection of performative utterances).

4   Ideological statements based on contents (what understanding of the relationship between oil, state, Petro-Socialism and culture do the speeches convey?).

In sum, the critical discourse analysis of public speeches elucidates the discursive strands that run across the collective meanings and ideologies that inform the conceptualisation of State Space, which manifests in the social practices of Bureaucratic Power that shape the formation of PDVSA's State Space and the entanglement between State Space, oil and culture to identify the performative utterances that indicate the agency, will and power to enact a new reality under Petro-Socialism.

Hugo Chávez's discursive construction of Petro-Socialism

Discourses can exert power because they are regulated, institutionalised and linked to action. In the case of Hugo Chávez, his discursive exercise of Bureaucratic Power was manifested in changes to reality through his embodiment of the New Magical State. The oil windfall that characterised the period that frames this book (2005–2013) provided Chávez with the resources to reform the institutional apparatus and to create new policy instruments to conceptualise the Socialist State Space. Considering the prolific quantity of political speeches and public broadcasts delivered by Chávez during his three presidencies, which would have made the selection of public speeches extremely laborious and time consuming, for the aims of this study, the selection is based on the collection of speeches included in the online archive of the former President of PDVSA, Rafael Ramírez, as relevant discursive events in matters of national oil policy.

The discursive plane from which Chávez spoke is framed by his embodiment of the New Magical State, the power of his performative utterances to enact a new reality manifested in the creation of new policy instruments and is echoed in the speeches of the leaders of PDVSA and PDVSA La Estancia and novel visual representations of Petro-Socialism in urban space, explored in Chapter 5.

The speeches analysed in this section were delivered during a period of steady rise in oil prices, they reached US$100 per barrel in 2008 but began a steady decline in 2013 to reach values below US$40 per barrel by early 2016 (OPEC, 2016). To select the discourse fragments in Hugo Chávez's speeches relevant for analysis, a word frequency query found that the five most frequently used words were:

Venezuela
Presidente (President)
Petróleo (Oil)
Desarrollo (Development)
Palabras (Words)

Discarding the common words Venezuela, President and Words, oil and development are the two most recurrent terms in the six speeches. Far behind, the terms *socialismo* and *socialista* are referenced 19 and 9 times respectively.

Socialista (socialist) appears only in two of the five speeches (Table 4.1).

Socialismo (socialism) is referenced in three of his speeches, most mentions concentrate in his presidential inauguration speech of January 2007 (Table 4.2).

Meanwhile, *petróleo* (oil) features in every speech but most predominantly in the speech delivered at the 141st Extraordinary Meeting of OPEC (2006), followed by the speech for the presentation of the Sowing Oil Plan (2005). In the remaining speeches, the term is mentioned 19 times (IV Summit of Petrocaribe and 3rd Summit of OPEC), to 5 and 3 (Inauguration Speech in 2007 and the LXI ONU General Assembly, respectively) (Table 4.3).

Table 4.1 Hugo Chávez speeches text query results

	References	Coverage (%)
Palabras del Presidente de la República, Comandante Hugo Chávez, 2007	8	0.08
Palabras del Jefe de Estado en la presentación del "Plan Siembra Petrolera", 2005	1	0.01

Table 4.2 Hugo Chávez speeches text query results

	References	Coverage (%)
Palabras del Presidente de la República, Comandante Hugo Chávez, 2007	16	0.15
Discurso del Presidente Hugo Chávez en la Sesión de Apertura de la 3era Cumbre de la OPEP, 2007	1	0.03
Palabras del Jefe de Estado en la presentación del Plan Siembra Petrolera, 2005	2	0.02

Table 4.3 Hugo Chávez speeches text query results

	References	Coverage (%)
Palabras del Presidente Chávez en la Instalación de la 141ª Reunión Extraordinaria de la OPEP, 2006	57	0.53
Palabras del Jefe de Estado en la presentación del "Plan Siembra Petrolera", 2005	43	0.41
Discurso del Presidente Hugo Chávez en la Sesión de Apertura de la 3era Cumbre de la OPEP, 2007	19	0.41
Palabras del Presidente Hugo Chávez Frías en la inauguración de la IV Cumbre de Petrocaribe, 2007	19	0.17
Palabras del Presidente de la República, Comandante Hugo Chávez, 2007	5	0.04
Hugo Chávez LXI Asamblea General de la ONU, 2007	3	0.05

To explore which kind of relationship between oil and socialism is inferred within Chávez's speeches a cross reference of *socialista, socialismo* and *petróleo* were used to identify and extract the relevant discourse fragments (all translations by the author) (Table 4.4).

The following statements were extracted from the discourse fragments by identifying the relevant performative utterances:

A **I propose in honour of the spirit of that phrase** [sowing oil] and why not, to that of doctor Uslar and that of Juan Pablo Pérez Alfonzo and to all those that warned, wrote, said, fought and even died for the national oil sovereignty, **that we call this 2005–2030 Plan, Sowing Oil**.

B **This project will be**, from today, one of the enclaves, of the levers **to carry forward the socialist project**, to not truss the capitalist model of exploitation.

C Now oil belongs to the Venezuelan people and **oil will be for justice, for equality, for the development of our people, that is the truth**.

D OPEC was born 45 years ago. OPEC was brought to its knees. OPEC arose. **OPEC will live with success for ever. So be it**.

E Nothing and nobody will divert us from the road towards the Bolivarian socialism, the **Venezuelan socialism, our socialism**.

F We are headed towards the **Socialist Republic of Venezuela**.

G Towards the conformation of a communal State and the old bourgeois State that is still alive, alive and kicking, **we have to dismount it progressively whilst we are building the communal State, the socialist State**, the Bolivarian State. (…) Transform the old counter revolutionary State into a revolutionary State.

Chávez's performative utterances around oil are neither founded in fiction nor are they mere dramaturgical exercises of Bureaucratic Power. Petro-Socialism relied on the certainty that Venezuela stands over the biggest proven oil reserves in the world, surpassing Saudi Arabia's (Rowling, 2012), which led Chávez to assume that Venezuela would never run out oil, and would enjoy a never-ending supply of oil rent. This meant that Petro-Socialism's merger of oil rentierism with socialism could endure forever, fed by the illusion of never-ending high oil revenues. This illusion is what informs the performative utterance in statement D: 'OPEC will live with success for ever. So be it'. The *so be it* transforms an expression of hope into truth, and thus Chávez reaffirms his power as the leading figure of the revolution and of OPEC.

These certainties are also reflected in Chávez's characterisation of Petro-Socialism: Venezuelan, Bolivarian and oil-based. The statement 'I propose (…) that we call this Plan 2005–2030, Sowing Oil', is evidence of the power of Chávez's discourse; he is the sole authority who can verbalise the performative utterance that gives birth and names the strategy envisioned to govern the Venezuelan oil industry, and the country at large, for 25 years: the Sowing Oil Plan. It was also a way of affirming he would succeed in 'sowing oil' where previous leaders had failed. Chávez's adoption of Uslar Pietri's farming language to

Table 4.4 Hugo Chávez speeches discourse fragments

Extractos del discurso ofrecido por el Ciudadano Presidente de la República Bolivariana de Venezuela, Hugo Chávez Frías, con motivo de la presentación de los Planes Estratégicos de Petróleos de Venezuela PDVSA (Presentation of PDVSA Strategic Plans), 18 August 2005, Caracas.	He said, like saying farewell at over 90 years of age: 'we could not or did not know how or did not want to sow oil'. I propose in honour to the spirit of that phrase and why not, to that of doctor Uslar Pietri and that of Juan Pérez Alfonzo and to all of those who warned, wrote, said, fought and even died for oil sovereignty, that we call this 2005–2030 Plan, Sowing Oil. Then this Project is going to be, from today, one of the enclaves, of the levers to carry forward the socialist project, to not truss the capitalist model of exploitation. That would be contrary to constitutional mandate and contrary to national interest, but nobody should be scared by this, it is about equality and the economic, social, integral development of the country.
Palabras del Presidente Chávez en la Instalación de la 141ª Reunión Extraordinaria de la Conferencia Ministerial de la Organización de Países Exportadores de Petróleo OPEP (141st Extraordinary meeting of the Ministerial Conference of the Organisation of Petroleum Exporting Countries), 1 June 2006, Caracas.	There never was one drop of oil for the people of Venezuela, oil was sucked by the creole oligarchy and above all by the North-American empire. Oil now belongs to the Venezuelan people and oil will be for justice, for equality, for the development of our people, that is the truth. OPEC was born 45 years ago. OPEC was put to its knees. OPEC arose. OPEC will live with success forever. So be it.
Palabras del Presidente de la República Comandante Hugo Chávez (first public speech of 2007, on the inauguration of Hugo Chávez third presidential period). 8 January 2007, Caracas.	Nothing and nobody will divert us from the road towards Bolivarian socialism, Venezuelan socialism, our socialism. Second motor: the socialist constitutional reform, we are moving towards the Socialist Republic of Venezuela and that requires a profound reform of the National Constitution, our Bolivarian Constitution. We have to march towards the formation of a Communal State and the oil Bourgeois State that still lives, that is still alive and kicking, we have to dismantle it progressively whilst we are building the Communal State, the Socialist State, the Bolivarian State. A State that has the conditions and the capacity to drive a revolution. Almost all states were born to halt revolutions. What a challenge we have! To turn the oil counter revolutionary state into a revolutionary State.

underpin that oil would finally be 'harvested' highlights that his Bureaucratic Power as the New Magical State (López-Maya, 2007; Joyce and Bennett, 2010; Coronil, 2011, p. 4) was tied to the land and its subsoil, where all the powers of nature and its resources originate.

The purpose of the Sowing Oil Plan is cemented in the performative utterances in statements B, E, F and G. At the time these speeches were delivered, the Socialist State had yet to be realised, but it is verbalised as a present truth. He also consolidates the Sowing Oil Plan as one of the pillars of Petro-Socialism, moulding the role of PDVSA as the engine of revolutionary change: oil put at the service of socialism (Maass, 2009, pp. 202, 215). He stresses the particular relationship his model established between oil, the state and the people in the performative utterance in statement C: 'oil now belongs to the Venezuelan people'. As discussed in Chapter 1, the Venezuelan Petrostate holds the monopoly over oil, and in turn, oil has historically mediated the relationship between society and the state. By embodying the New Magical State, Chávez was able to construct a direct relationship between oil and the people while exerting full control over the distribution of oil revenue. The reality of the statement is enacted by adding that 'oil will be for justice, for equality, for the development of our people' and if any doubts lingered in his audience, he closes by declaring that 'that is the truth'. Although Chávez's proposal to reform the Constitution was defeated in a referendum in 2007, he proceeded to carry out the proposed reforms by governing by decree to devise the policy instruments for the dismantlement of what he characterised as the 'counter revolutionary state'.

Chávez's anti-capitalist and anti-neoliberal model for the Petro-Socialist State would never cease to be an oil rentier state. While he discursively called for the eradication of capitalism, he also advocated for the longevity of OPEC and global oil capitalism. Given that the vast majority of state revenue comes from oil exports, it highlights an intrinsic contradiction of Petro-Socialism. However, Chávez reconciled this incongruence when he affirmed that while in the past oil was used as a force of capitalist domination and plunder by oil corporations and a neoliberal PDVSA loyal to the Empire (United States of America), in Petro-Socialism oil would become a force of liberation, equality and development.

PDVSA's Sowing Oil Plan was devised as an alternative model to the neoliberal strategies of the Oil Opening Plan of the 'old' PDVSA. The name uses farming language to underpin the idea that oil can be 'sowed' to suggest a natural renewable cycle of sowing and harvesting the subsoil for oil, which as will be discussed later, informs the notion of Culture as Renewable Oil construed by PDVSA La Estancia. Thus, the advancement of Petro-Socialism called for the dismantlement of the 'old' institutional structures to pave the way for the consolidation of the socialist State.

Chávez did not envision a post-oil world, on the contrary, his model relied on the certainty of standing over the largest crude oil reserves in the world, a potentially inexhaustible supply of oil rent that assured the endurance of

Petro-Socialism and the Socialist State; a world in which OPEC would live on forever to guarantee high oil prices and a continuous flow of oil rent to the state's coffers. For Chávez, oil rentierism was necessary for the irreversibility of socialism. The policies and strategies created for the advancement of Petro-Socialism stood on unrealistic expectations of inexhaustible high oil revenues, that began to decline before his death in 2013.

Rafael Ramírez and the new PDVSA as agents of Petro-Socialism

Engineer Rafael Ramírez, President of PDVSA between 2004 and 2013, was one of Hugo Chávez's closest allies. During his tenure, Ramírez occupied simultaneously the posts of president of PDVSA, the Minister of Energy and Petroleum (2002–2014), Vice-President of Territorial Development and Vice-President of the United Socialist Party of Venezuela (PSUV) (Párraga, 2010, p. 122; Colgan, 2013, pp. 205–206). Effectively, Ramírez had almost complete control over the national oil industry, his discourse plane is that of an elite with Bureaucratic Power over the oil industry and sway over national politics, establishing the non-transferability of his performative utterances.

Following Chávez's mandate, Ramírez delineated PDVSA's new socialist profile (Párraga, 2010, pp. 24–26) expanding its functions beyond its core commercial mission of producing maximum oil revenue to the state by channelling the investment of the oil rent into the construction of the Socialist State. Ramírez discourse is able to exert power and construct a new national reality because he is in complete control of the entity that extracts and commercialises oil as well as the bureaucratic entities that govern its distribution. The wealth produced by PDVSA's participation in the global oil market is put at the service of the New Magical State and Petro-Socialism. This section traces the reverberations of Chávez's discourses within the public speeches of Rafael Ramírez.

The five most frequently used words by Ramírez are:

PDVSA
Petróleo (Oil)
Petrolera (Oil)
Nuestro/Nuestros (Our/Ours)
Nacional (National).

Unsurprisingly, the most referenced terms are the company itself and oil; the terms *nuestro/nuestros* (our/ours) and *nacional* (national) indicate a nationalist slant in regards to oil in tune with Hugo Chávez's rhetoric. 'Socialism' is referenced in five of his speeches between 2012 and 2014, most mentions concentrated on his speech delivered to celebrate the centenary of the burst of the first oil field Zumaque I in 2014. 'Socialist' appears only on three of his 16 speeches delivered between 2012 and 2014, most notably in the speech delivered at the hearing of Ministers for Territorial Development in November 2013 (eight months after Chávez's death, seven months into Nicolás Maduro's presidency).

Going back to the beginnings of Chávez's second presidency and the launch of the PPS, it is noticeable that socialist and socialism are absent from Ramírez's speech delivered in August 2005 to launch the Sowing Oil Plan. It is mentioned only once, in the speech delivered to the National Assembly in May of that same year, in the last line of the last paragraph:

> The new PDVSA has a beautiful role: contributing in an effective manner, with all our capacity, in liberating our people from the shadows of misery, of exclusion. The new PDVSA is an instrument of its people, of its revolution to build new economic-social relations, that support and foster the high degree of conscience and mobilisation achieved by our people through all the battles fought in defence of the revolution: the defeat of the coup d'état, the defeat of the oil sabotage, the defeat of violence and destabilisation, the victory of Santa Inés battle and many more battles we have yet to fight in pursuit of the construction of a homeland of free men, where the principles of solidarity and love lead inevitably to the happiness of our people, *to socialism*.
>
> (Ramírez, 2005, p. 16, translation and emphasis by the author)

To trace the discursive entanglements between oil and socialism, a cross reference query of the terms 'socialist', 'socialism', 'oil' and 'sowing oil' identified discourse fragments that intersect oil, PDVSA's Sowing Oil Plan and Socialism, in chronological order (Table 4.5).

The following statements were extracted from the discourse fragments for analysis, identifying relevant performative utterances:

H   It is a revolutionary policy because that oil income is going to become (…) the main lever to create the material foundations for the **construction of socialism in our country**.

I   (…) our revolution is profoundly anti-imperialist and **socialism is the path of our salvation**. It is the socialist fatherland or death.

J   The **oil rent** must be an instrument for the construction of a new economic order, it **must be an instrument for the construction of socialism**.

K   The society our Eternal Commander **Hugo Chávez dreamt: 'the socialist society'**.

The performative utterance in statement H establishes PDVSA's new identity as a revolutionary institution. By defining the Sowing Oil Plan as a revolutionary policy, he mirrors Chávez's assertion that the national oil policy laid the foundations for the advancement of Petro-Socialism and the construction of the Socialist State. Here Ramírez declares that his Bureaucratic Power as president of the state-owned oil company, Minister of Oil and Energy and Vice-President of the

Table 4.5 Rafael Ramírez speeches discourse fragments

Speech by the Minister of Popular Power of Oil and Mining, Rafael Ramírez, for the presentation of PDVSA's Annual Report 2012. Caracas, 3 May 2013.	It is a revolutionary policy because oil income is going to become and that is how it is established in our Plan of the Fatherland, the main lever to overcome the oil rentierist model, the main lever to create the material foundations for the construction of socialism in our country.
PDVSA Loyal to CHÁVEZ Legacy, speech by Minister of Popular Power of Oil and Mining and President of Petróleos de Venezuela, S. A., Rafael Ramírez, for the PDVSA Productive National Sector Session. Maracaibo, Zulia, 16 May 2013.	We intend to capture that vast oil rent so we can build an alternative model to the oil rentier model, a productive model that we have said, will be socialist.
Hearing of Report and Accounts of the Ministers of the Vicepresidency of the Area of Territorial Development. Caracas, 4 November 2013.	President Chávez, indisputable leader of our revolutionary process, has said it with clarity and courage, our revolution is profoundly anti-imperialist and socialism is the path of our salvation. It is the socialist fatherland or death.
Special Session for the Centenary of the beginnings of the commercial activity of the oil industry, with the exploitation of Zumaque I oil well. Maracaibo, 5 August 2014.	The oil rent must be an instrument for the construction of a new economic order, it must be an instrument for the construction of socialism.
	The society dreamt by our Eternal Commander Hugo Chávez: 'the socialist society', is the best recognition we can make to the selfless, disinterested work, to the love of the humble, to the Fatherland of Commander Chávez.

ruling party, is subservient to Chávez. His discourse is able to exert power and contribute to construct the Socialist State because it is institutionalised and linked to action. The Sowing Oil Plan's 25 years timeframe is an instrument put at the service of the power of the New Magical State embodied by Hugo Chávez, to contribute in the construction and perpetuation of Petro-Socialism. Along these lines, echoes of Chávez's discourse on the Socialist State are found in statements B, C, D and E. Ramírez's performative utterances do not describe reality, they describe the act of creation of a future that is built as it is verbalised, in which the oil rent is presented as the foundation to build the Socialist State and the future socialist society.

However, it is contradictory for the head of an international oil corporation to affirm that the path of salvation from oil rentierism and capitalism is found only in socialism as PDVSA's *raison d'etre* is to produce the highest oil revenue to the state by taking part in the dynamics of global oil capitalism. There is a conceptual conflict in Ramírez's condemnation of capitalism: the oil rent that responds to the capitalist economic global order is put at the service of creating an alternative economic model to capitalism within the national territory, which is what Petro-Socialism was meant to achieve. Overall, Ramírez's performative utterances conceal that Venezuelan oil revenues are indeed enmeshed with global capitalism and highly dependent on the fluctuations of international oil markets. Nonetheless, the purpose of his speeches is to discursively establish PDVSA's identity as a revolutionary oil corporation. Hence, the relationship between oil and the state that Ramírez speeches convey reaffirms PDVSA's allegiance to the revolution, he instrumentalises his share of Bureaucratic Power as the head of the state-owned oil company to contribute to Chávez's vision.

PDVSA La Estancia, an instrument of the Sowing Oil Plan that 'harvests culture'

Venezuelan Lawyer Beatrice Sansó de Ramírez, spouse of Rafael Ramírez, occupied the post of General Manager of PDVSA La Estancia between 2005 and 2015; she had previously served as legal advisor to PDVSA. Her discourse plane is defined as an elite with a share of Bureaucratic Power within and outside the oil industry which establishes the non-transferability of her performative utterances.

This section examines the speeches of Beatrice Sansó, the General Manager of PDVSA La Estancia, to trace the discursive entanglements between Petro-Socialism, the Sowing Oil Plan and culture, which frames the institution's conceptualisation of the city as an oil field and its use of language to define PDVSA La Estancia as 'oil that harvests culture'. This discursive construction justifies PDVSA La Estancia's General Manager to interpret the Organic Law of Hydrocarbons as an implicit cultural policy. The legal vacuum in territorial ordnance left between 2005 and 2010 enabled PDVSA to construct a parallel State Space through the Oil Social Districts, and for PDVSA La

Estancia to instrumentalise the Organic Law of Hydrocarbons as an implicit cultural policy in which culture is conceived in Yúdice's terms of the expediency of Culture as a Resource (2003), reframed within the extractive logic of the oil industry.

A word frequency query of all speeches found that the five most frequently used words are:

Estancia
PDVSA
Caracas
Paraguaná
Maracaibo

Although PDVSA La Estancia has a national scope, its works and discourse are predominantly concentrated in Caracas. Text search queries of *cultura* (culture), *cultural* (cultural), *petróleo* (oil), *petrolero* (oil), *socialista* (socialist), *socialismo* (socialism) and *siembra petrolera* (sowing oil) were conducted to identify discourse fragments that intersect oil, culture and socialism, as well as the work of PDVSA La Estancia with the Sowing Oil Plan. The discourse fragments identified are collated here (Table 4.6).

The following performative utterances have been extracted from the discourse fragments above:

L … identify through our work the social and **cultural fruits of oil.**
M … substitute it for the **Utopia of the Possible**, ideal that inspires the daily actions of the **Cultural Oasis** of the capital …
N a **socialist city**, paradigm of the city of the twenty-first century.
O PDVSA-Centro de Arte La Estancia has given **a radical turn to cultural management** in regards to a practical and unique conception of our policies, in accordance with **Article 5 of the Organic Law of Hydrocarbons** and the **Sowing Oil Plan** …
P … where **Sowing Oil** will find once again furrows to inseminate itself, achieving the **utopia of the possible**.
Q Its actions are based on the principles of the **Sowing Oil Plan**/dedicated to this **sowing oil** of PDVSA La Estancia/PDVSA La Estancia, **instrument of the Sowing Oil Plan**/this instrument of **Sowing Oil** that belongs to us all.
R PDVSA La Estancia: **Utopia of the Possible.**
S PDVSA La Estancia **Oil that harvests culture.**

Throughout her speeches, Sansó de Ramírez threads a discursive strand that joins together the Sowing Oil Plan and the cultural work of PDVSA La Estancia (statements collected in Q), in particular the restoration of public art and regeneration of public space in the cities of Caracas, Maracaibo and Paraguaná. Sansó's performative utterance that states that PDVSA La Estancia 'has given a radical turn to cultural management' enacts the way it functions as the cultural arm of the state-owned oil company, setting it apart from any other national or local cultural institution in Venezuela: it has direct and discreet access to the oil rent. Its policies are aligned with the Sowing Oil Plan and subordinate to the Organic Law of Hydrocarbons, more specifically with Article 5. At this point it is worth quoting in full Article 5, Chapter II Activities Related to Hydrocarbons:

> Article 5. The activities regulated under this Law shall be directed at promoting a comprehensive, organic and sustained development of the country, focused on a rational use of this resource and the preservation of the environment. To this end the strengthening of the national productive sector and the processing of raw materials produced by hydrocarbons in the country shall be promoted, as well as the incorporation of advanced technologies. The revenues that accrue from hydrocarbons to the nation shall contribute to the funding of healthcare, education, the creation of macroeconomic stabilisation funds and productive investment, to achieve an appropriate link between oil and the national economy, *all of this in the service of the wellbeing of the people.*
>
> (Asamblea Nacional de la República Bolivariana de Venezuela, 2006, pp. 1–2, translation and emphasis by the author)

Although Article 5 does not make any explicit reference to culture (or cultural activities) it is implicit in the focus on social development expressed as 'in the service of the wellbeing of the people', here wellbeing is enclosed under the socio-cultural branch of the Sowing Oil Plan, in which PDVSA La Estancia is the social and cultural extension of PDVSA:

> The objectives of the Sowing Oil Plan were directed to two different branches: *The first* is the *socio-economic* sphere, referring to the fulfilment of social or economic operation; and the *second* to the *socio-cultural* sphere, through the conformation of an organism that operates as the socio-cultural arm of PDVSA.
>
> (Rondón de Sansó, 2008, p. 417, translation by the author, emphasis in the original)

PDVSA La Estancia fulfils the socio-cultural sphere of the Sowing Oil Plan by invoking Article 5 of the Organic Law of Hydrocarbons as an implicit cultural policy in the terms defined by Jeremy Ahearne: 'any political strategy that looks to work on the culture of the territory over which it presides' (Ahearne, 2009, pp. 143–144). According to the Sowing Oil Plan, PDVSA La Estancia's objectives are to 'promote and disseminate the historic and artistic heritage of the country, including the restoration of the fundamental works that conform that heritage' (Rondón de Sansó, 2008, p. 425), all encompassed within the Petro-Socialist State Space that PDVSA presides over. PDVSA La Estancia establishes

Table 4.6 Beatrice Sansó Ramírez speeches discourse fragments

PDVSA La Estancia Arte para todos (PDVSA La Estancia Art for all), PDVSA LA Estancia Website, 2005.	… we want people to go on the street and say 'that was done by Centro de Arte La Estancia', 'La Estancia put its hand there', so that they identify through our work the social and **cultural fruits of oil.** … Its actions are based on the principles of the **Sowing Oil Plan.**
Discurso con motivo de la develación del Abra Solar de Alejandro Otero (Speech on the occasion of Unveiling of Alejandro Otero's Abra Solar). Caracas, 9 November 2007.	…we will eradicate once and for all the terrible pessimism of the impossible and substitute it for the **Utopia of the Possible,** ideal that inspires the daily actions of the **Cultural Oasis** of the capital, and that we wish to project as a model of action that allows to reach the goal of living in a beautiful Caracas, with all its spaces recovered, a **socialist city,** paradigm of the city of the twenty-first century.
Discurso con motivo de la develación de la Fisicromía del maestro Carlos Cruz Diez (Speech on the occasion of Unveiling maestro Carlos Cruz Diez's Fisicromía). Caracas, 13 March 2008.	… to end once and for all with the terrible pessimism of the impossible to substitute it for the **Utopia of the Possible,** an ideal that inspires the daily actions of this **Cultural Oasis,** and that we wish to project as a model of action that allows to achieve by the year 2010, on the 19 April Bicentenary, the goal of living in a beautiful Caracas, with all its spaces recovered, a **socialist city,** paradigm of the city of the twenty-first century.
Discurso con motivo de la inauguración de la exposición Mateo Manaure: el Hombre y el Artista (Speech on the occasion of Opening the exhibition Mateo Manaure: the Man and the Artist). Caracas, 17 July 2008.	… paradigm of the possible, **utopia** of the desirable …
Discurso con motivo de la Inauguración de PDVSA La Estancia Paraguaná (Speech on the occasion of Opening PDVSA La Estancia in Paraguaná). Paraguaná, 13 August 2008.	The '**Sowing Oil Plan**' through its national, popular and revolutionary strategy, immerses itself into the entrails of the land, to pump and nurture the heart of an effervescent Venezuela This way, PDVSA-Centro de Arte La Estancia has given **a radical turn to cultural management** in regards to a practical and unique conception of our policies, in accordance with **Article 5** of the Organic Law of Hydrocarbons and the **Sowing Oil Plan** … I did not want to finish without reciting the singer of the struggle, Alí Primera, who recites Paraguaná … and we do it because today, Alí, we are certain, if he was alive, as a necessary song, he would write new lyrics, dedicated to this **sowing oil** of PDVSA La Estancia.

Inauguración de la sede de PDVSA La Estancia Maracaibo (Opening of PDVSA La Estancia branch in Maracaibo). Maracaibo, 5 February 2010.	... a space for music and Venezuelanness; a space for dance, for big and small children, where **Sowing Oil** will find once again furrows to inseminate itself, achieving the **utopia of the possible.**
A un año de labor PDVSA La Estancia Petróleo que cosecha cultura (After a year of labour **PDVSA La Estancia Oil that harvests culture**). Maracaibo, 5 February 2011.	**'Oil that harvests culture'** enhances our already beautiful central courtyard, with a photographic journey through the work of PDVSA La Estancia Maracaibo, during 2010. Over 560 activities between concerts, workshops, social seminars, film, theatre, dance and children programming entirely free. We keep finding for **Sowing Oil**, the furrows to inseminate it, and continue, inspiring the possible!
Palabras como oradora de orden en la Sesión Solemne del Concejo Municipal del Municipio Carirubana, Estado Falcón (Speaker of the Solemn Session of the Municipal Council of Carirubana Municipality, Falcón State). 8 March 2012.	Today, PDVSA LA ESTANCIA, **instrument of the Sowing Oil Plan** blooms in the faces of boys and girls replacing their audio visual pale gaze for the smile of playing under the sun; irrigates spaces with the colours of the tropic and invokes Alí Primera through its work because we are certain, as we have said in our opening speech of PDVSA LA ESTANCIA Paraguaná that as a necessary song would have dedicated its lyrics to **Sowing Oil.**
PDVSA La Estancia: La Utopia De Lo Posible (PDVSA La Estancia: Utopia of the Possible). Caracas, 7 May 2012.	PDVSA La Estancia: **Utopia of the Possible** ... nurture the urge to move forward of those of us who have had the fortune of having this instrument of **Sowing Oil** that belongs to us all.

itself as a national management entity bound to the Organic Law of Hydrocarbons, which in their view prevails over regional and municipal laws, giving them the authority to govern over the space of Caracas, as if it were a quasi-state authority. 2007 was also the year in which PDVSA La Estancia extended its scope of action beyond an arts centre, through its nationwide expansion establishing branches in Paraguaná and Maracaibo, cultural programming across the country, and most notably by embarking on an ambitious programme of regeneration of public art and public spaces, effectively taking over the functions of local government institutions and municipalities, claiming ownership over Caracas. Consequently, PDVSA La Estancia presents its actions as a form of sowing oil, bearing the 'social and cultural fruits of oil' through their work in the city. Sansó is aiming for a direct identification between oil, people and PDVSA La Estancia, as well as a direct correlation between oil and culture.

The continued reference to Uslar Pietri's 'to sow the oil' is used to claim, as in the performative utterance of statement P, that they are opening furrows to scatter and plant 'seeds of oil' that grow into an 'utopia of the possible', establishing that only through culture can oil be sown, relating culture back to the territory. Sansó discursively engages oil and culture in a symbiotic and cyclical farming relationship. PDVSA La Estancia is a form of sowing oil, and once sown, culture is 'harvested' and therefore PDVSA La Estancia becomes 'oil that harvests culture', an entity that transforms a non-renewable mineral resource into a renewable resource, materialised in their programme of cultural events, and most poignantly through public art restoration and urban regeneration. Sansó's vision of culture, to paraphrase Bauman (2004, p. 64), is seen through the eyes of the farmer-manager whose growing field is in this case the city as an oil field, where culture is 'extracted' from as if it was a mineral deposit accumulated in the subsoil, the basis of the proposition of the notion of Culture as Renewable Oil.

Caracas is the location of PDVSA's headquarters, the 'Cultural Oasis' of the capital, and where most of its works of restoration and urban regeneration have taken place. The beautification of Caracas aims to make it a performative enactment of the future reality of the Petro-Socialist city of the twenty-first century: 'Caracas beautiful, with all its spaces recovered, a socialist city, paradigm of the city of the twenty-first century'. In practice, the emphasis was put on the restoration of public art and public spaces in Caracas rather than building anew. The focus on Caracas is intentional. The majority of the public art and public spaces restored are the legacy of the golden years of the Magical State, symbols of the modern oil nation, prior to the presidencies of Hugo Chávez. The socialist city, or more specifically, the Petro-Socialist city, manifests as the recovery of the city as it stood during the decades of the previous oil boom in contrast to the extreme deterioration the city fell under in the 1990s, exacerbated during the first five years of Chávez's presidency. In this sense, the performative utterance in statement L – 'identify the social and cultural fruits of oil through our work' – indicates a will to repossess these symbols and mediate people's perceptions of oil through the interventions of PDVSA La Estancia.

This is aligned with a discursive strand reiterated to the letter in different speeches delivered between 2007 and 2012: *utopía de lo posible* (utopia of the possible). Beatrice Sansó does not present utopia as an unattainable ideal world in the future. On the contrary, it is 'possible', attainable and realised in the present by PDVSA La Estancia's 'sowing oil' and 'harvesting culture' throughout the spaces of the city. The inspirational language presents the state-owned oil company as the producer of the material spaces of an oil-based utopia, a realised utopia, meaning that when they sow the oil, they harvest it as culture, and once harvested, the Petro-Socialist utopian city becomes 'real'. But this Petro-Socialist city is in essence an oil city/oil field, owned, cared for and managed by PDVSA and its 'utopian' oil workers. The once deteriorated spaces of the city are equated to capitalist decay compounded by a neoliberal PDVSA, while the 'Caracas beautiful' discursively reframes the areas recovered by PDVSA La Estancia as the model of the future Petro-Socialist city as they are represented in the adverts of the campaign 'We transform oil into renewable resource for you' analysed in Chapter 5.

Conclusion

Petro-Socialism is an oil boom phenomenon, it emerged at a moment when oil prices were at an all-time high. PDVSA was central to guarantee the success of Petro-Socialism which relied on the certainty that Venezuela sits over the biggest proven crude oil reserves in the world leading Chávez to assume that Venezuela would enjoy a never-ending supply of high oil revenue. This contradiction runs across State Space and Bureaucratic Power particularly through the instrumentalisation of culture to construct the New Magical State's illusion of Culture as Renewable Oil; the recurrent use of the phrases 'sowing oil, 'harvest culture' and 'utopia of the possible' are grounded on this discursive construction. As an instrument of the Sowing Oil Plan, the act of sowing oil to harvest culture is an illusion of the New Magical State: oil ceases to be finite when it is sown to bear the fruits of culture. Within this vision, the material space of the 'petro-socialist city' is meant to be absorbed by the Oil Social District and populated by people directly identified with oil. Beatrice Sansó's discursive use of culture to frame the institution's actions as a 'utopia of the possible', are based on construing the city as an oil field while conceptualising a symbiotic and cyclical relationship between oil, land and culture condensed in the notion of Culture as Renewable Oil by stating that PDVSA La Estancia is 'oil that harvests culture'.

References

Ahearne, J. (2009) 'Cultural Policy Explicit and Implicit: A Distinction and Some Uses', *International Journal of Cultural Policy*, 15(2), pp. 141–153.
Asamblea Nacional de la República Bolivariana de Venezuela (2006) *Ley Orgánica de Hidrocarburos*. Caracas, Venezuela: Asamblea Nacional de la República Bolivariana

de Venezuela. Available at: www.pdvsa.com/images/pdf/marcolegal/LEY_ORGANICA_ DE_HIDROCARBUROS.pdf (accessed: 17 July 2018).

Bauman, Z. (2004) 'Culture and Management', *Parallax*, 10(2), pp. 63–72.

Colgan, J. (2013) *Petro-Aggression When Oil Causes War*. Cambridge: Cambridge University Press.

Coronil, F. (1997) *The Magical State: Nature, Money, and Modernity in Venezuela*. Chicago: University of Chicago Press.

Coronil, F. (2011) 'Magical History What's Left of Chávez?', in *LLILAS Conference Proceedings, Teresa Lozano Long Institute of Latin American Studies*. Latin American Network Information Center, Etext Collection.

Cuddon, J. A. (2013) *A Dictionary of Literary Terms and Literary Theory*, 5th edn. Chichester, West Sussex, UK: Wiley-Blackwell.

Fairclough, N. (2010) *Critical Discourse Analysis: The Critical Study of Language*, 2nd edn. Harlow: Longman.

Grauer, O. (2015) 'Interview February 6th 2015'. Boston, Massachusetts: interview.

Jacobsen, K. (1971) 'How to Make the Distinction Between Constative and Performative Utterances', *The Philosophical Quarterly*, 21(85), pp. 357–360.

Jager, S. (2011) 'Discourse and Knowledge: Theoretical and Methodological Aspects of a Critical Discourse and Dispositive Analysis', in Meyer, M. and Wodak, R. (eds) *Methods of Critical Discourse Analysis*. London: Sage Publications Ltd, pp. 32–62.

Joyce, P. and Bennett, T. (eds) (2010) *Material Powers: Cultural Studies, History and the Material Turn*. London: Routledge.

López-Maya, M. (2007) *Nuevo debut del Estado mágico, Aporrea.org*. Caracas, Venezuela. Available at: www.aporrea.org/actualidad/a35326.html (accessed: 8 December 2015).

Maass, P. (2009) *Crude World*. London: Allen Lane.

OPEC (2016) *OPEC Basket Price*. Available at: www.opec.org/opec_web/en/data_ graphs/40.htm (accessed: 15 April 2016).

Párraga, M. (2010) *Oro Rojo*. Caracas: Ediciones Puntocero.

PDVSA La Estancia (no date a) *Historia*. Available at: www.pdvsalaestancia.com/?page_ id=131 (accessed: 14 April 2016).

PDVSA La Estancia (no date b) *Zonas de Injerencia*. Available at: www.pdvsalaestancia. com/?page_id=574 (accessed: 14 April 2016).

Pérez Alfonzo, J. P. (2011) *Hundiéndonos en el excremento del diablo*. Caracas, Venezuela: Banco Central de Venezuela.

Pérez Schael, M. S. (1993) *Petróleo, cultura y poder en Venezuela*. Caracas, Venezuela: El Nacional.

Pinardi, S. (2007) 'Las misiones y el discurso espectáculo', in *II Jornadas de la sección de Estudios Venezolanos, Universidad Central de Venezuela*. Caracas, Venezuela: Latin American Studies Association.

Quintero, R. (2011) 'La cultura del petróleo: ensayo sobre estilos de vida de grupos sociales de Venezuela', *Revista BCV*, pp. 15–81.

Ramírez, R. (2005) *Discurso de Rafael Ramírez, Presidente de PDVSA y Ministro de Energía y Petróleo, ante la Asamblea Nacional, 25 de Mayo de 2005*. Caracas, Venezuela. Available at: www.aporrea.org/energia/a14338.html (accessed: 12 August 2015).

Rondón de Sansó, H. (2008) *El régimen jurídico de los hidrocarburos: el impacto del petróleo en Venezuela*. Caracas, Venezuela: Epsilon.

Rowling, R. (2012) *Venezuela Passes Saudis to Hold World's Biggest Oil Reserves*, *Bloomberg*. Available at: www.bloomberg.com/news/articles/2012-06-13/venezuela-overtakes-saudis-for-largest-oil-reserves-bp-says-1- (accessed: 15 April 2016).

Sato Kotani, A. (2015) *Recuerdos del futuro*, El Nacional. Caracas, Venezuela. Available at: www.el-nacional.com/papel_literario/Recuerdos-futuro_0_616738423.html (accessed: 14 April 2016).

Thompson, J. B. (1984) *Studies in the Theory of Ideology*. Berkeley, CA: Polity.

Tinker Salas, M. (2014) *Una herencia que perdura, petróleo, cultura y sociedad en Venezuela*. Caracas, Venezuela: Editorial Galac.

Yúdice, G. (2003) *The Expediency of Culture: Uses of Culture in the Global Era*. London: Duke University Press.

Speeches

Chávez, Hugo (2005) *Extractos del discurso ofrecido por el Ciudadano Presidente de la República Bolivariana de Venezuela, Hugo Chávez Frías, con motivo de la presentación de los Planes Estratégicos de Petróleos de Venezuela (PDVSA)*, 18 August, Hotel Caracas Hilton, Caracas. Available at: http://rafaelramirez.desarrollo.org.ve/ (accessed: 12 August 2015).

Chávez, Hugo (2006) *Palabras del Presidente de la República Bolivariana de Venezuela Hugo Chávez, en la Instalación de la 141ª. Reunión Extraordinaria de la Conferencia Ministerial de la Organización de Países Exportadores de Petróleo (OPEP)*, 1 June, Caracas. Available at: http://rafaelramirez.desarrollo.org.ve/ (accessed: 12 August 2015).

Chávez, Hugo (2006) *Intervención del presidente de la República Bolivariana de Venezuela, Hugo Chávez, en la LXI Asamblea General de la Organización de las Naciones Unidas (ONU)*, 20 September, United Nations Headquarters, New York. Available at: http://rafaelramirez.desarrollo.org.ve/ (accessed: 12 August 2015).

Chávez, Hugo (2007) *Intervención del Ciudadano Presidente de la República Bolivariana de Venezuela, Comandante Hugo Rafael Chávez Frías, en la Sesión de Apertura de la Tercera Cumbre de Jefes de Estado y de Gobierno de la Organización de Países Exportadores de Petróleo (OPEP)*, 17 November, Riad, Saudi Arabia. Available at: http://rafaelramirez.desarrollo.org.ve/ (accessed: 12 August 2015).

Chávez, Hugo (2007) *Palabras del Presidente de la República, Comandante Hugo Chávez*, 8 January, Caracas. Available at: http://rafaelramirez.desarrollo.org.ve/ articulos/palabras-del-presidente-de-la-republica-comandante-hugo-Chávez/ (accessed: 12 August 2015).

Chávez, Hugo (2007) *Palabras del Presidente Hugo Chávez Frías en la inauguración de la IV Cumbre de Petrocaribe*, 21 December, Cienfuegos, Cuba. Available at: http:// rafaelramirez.desarrollo.org.ve/ (accessed: 12 August 2015).

Ramírez Carreño, Rafael Darío (2005) *Palabras del Ministro Rafael Ramírez en la presentación del Plan Siembra Petrolera, extractos de la Conferencia Planes Estratégicos de PDVSA, presentados por el Ministro de Energía y Petróleo y Presidente de PDVSA, Rafael Ramírez Carreño*, 19 August, Hotel Caracas Hilton, Caracas. Available at: http://rafaelramirez.desarrollo.org.ve/articulos/palabrasdelministrorafaelramirezenla presentaciondelplansiembrapetrolera/ (accessed: 12 August 2015).

Ramírez Carreño, Rafael Darío (2005) *Discurso del ministro de Energía y Petróleo y presidente de PDVSA, Rafael Ramírez, con motivo del 45 aniversario de OPEP*, 14 September, Caracas. Available at: http://rafaelramirez.desarrollo.org.ve/articulos/ discurso-del-ministro-de-energia-y-petroleo-y-presidente-de-pdvsa-rafael-ramirez-con-motivo-del-45-aniversario-de-opep/ (accessed: 12 August 2015).

Ramírez Carreño, Rafael Darío (2005) *Discurso del Ministro de Energía y Petróleo y Presidente de PDVSA, Rafael Ramírez Carreño, en ocasión del Ciclo de Conferencias 'Venezuela, Política y Petróleo' en el marco del 45 aniversario de la Organización de Países Exportadores de Petróleo (OPEP), organizado por la Asociación Civil Casa Amarilla Patrimonio de Todos y el Instituto de Altos Estudios Diplomáticos Pedro Gual, del Ministerio de Relaciones Exteriores de la República Bolivariana de Venezuela*, 20 October, Caracas. Available at: www.pdvsa.com/interface.sp/database/fichero/publicacion_opep/2376/169.PDF (accessed: 3August 2015).

Ramírez Carreño, Rafael Darío (2005) *Discurso de Rafael Ramírez, Presidente de PDVSA y Ministro de Energía y Petróleo, ante la Asamblea Nacional*, 25 May, National Assembly, Caracas. Available at: www.aporrea.org/imprime/a14338.html (accessed: 12 August 2015).

Ramírez Carreño, Rafael Darío (2005) *Palabras del ministro de Energía y Petróleo y presidente de PDVSA, Rafael Ramírez, en el encuentro de fin de año con los periodistas nacionales e internacionales*, 21 November, Caracas. Available at: http://rafaelramirez.desarrollo.org.ve/articulos/palabras-del-ministro-de-energia-y-petroleo-y-presidente-de-pdvsa-rafael-ramirez-en-el-encuentro-de-fin-de-ano-con-los-periodistas-nacionales-e-internacionales/ (accessed: 12 August 2015).

Ramírez Carreño, Rafael Darío (2006) *Discurso del ministro de Energía y Petróleo y presidente de PDVSA, Rafael Ramírez, en el Tercer Seminario Internacional de la OPEP: 'OPEP en una nueva era energética: desafíos y portunidades'*, 21 September, Vienna, Austria. Available at: http://rafaelramirez.desarrollo.org.ve/articulos/plena soberaniapetrolera/ (accessed: 12 August 2015).

Ramírez Carreño, Rafael Darío (2006) *Discurso del ministro de Energía y Petróleo de Venezuela, Rafael Ramírez, en la Instalación de la 141a Reunión Extraordinaria de la OPEP*, 01 June, Caracas. Available at: http://rafaelramirez.desarrollo.org.ve/articulos/discurso-del-ministro-de-energia-y-petroleo-de-venezuela-rafael-ramirez-en-la-instalacion-de-la-141a-reunion-extraordinaria-de-la-opep/ (accessed: 12 August 2015).

Ramírez Carreño, Rafael Darío (2006) *Discurso del Ingeniero Rafael Ramírez, ministro de Energía Petróleo y presidente PDVSA, en la Exposición LatinoAmericana del Petróleo*, 27 June, Maracaibo. Available at: http://rafaelramirez.desarrollo.org.ve/?s= Discurso+del+Ingeniero+Rafael+Ram%C3%ADrez%2C+ministro+de+Energ%C3% ADa+Petr%C3%B3leo+y+presidente+PDVSA%2C+en+la+Exposici%C3%B3n+Latino Americana+del+Petr%C3%B3leo&x=0&y=0 (accessed: 12 August 2015).

Ramírez Carreño, Rafael Darío (2007) *Discurso del Ministro Ramírez durante el Foro Energético de Lisboa 2007*, 2 October, Lisbon, Portugal. Available at: http://rafael ramirez.desarrollo.org.ve/articulos/discursodelministroramirezduranteelforoenergetico delisboa2007/ (accessed: 12 August 2015).

Ramírez Carreño, Rafael Darío (2008) *Palabras del ministro Ramírez en encuentro con los medios de comunicación*, 18 March, PDVSA, Caracas, Venezuela. Available at: http://rafaelramirez.desarrollo.org.ve/articulos/palabrasdelministroramirezenencuentro conlosmediosdecomunicacion/ (accessed: 12 August 2015).

Ramírez Carreño, Rafael Darío (2012) *Discurso del ministro del Poder Popular de Petróleo y Minería y presidente de PDVSA, Rafael Ramírez, en la presentación del Informe de Gestión Anual de PDVSA 2012*, 3 May, PDVSA, Caracas, Venezuela. Available at: http://rafaelramirez.desarrollo.org.ve/articulos/discursodelministrorafaelramirezenla presentaciondelinformedegestionanualdepdvsa2012/ (accessed: 12 August 2015).

Ramírez Carreño, Rafael Darío (2013) *PDVSA Fiel al Legado de Chávez. Discurso del Ministro del Poder Popular de Petróleo y Minería y Presidente de Petróleos de Venezuela,*

S. A., Rafael Ramírez, en la Jornada PDVSA Sector productivo nacional conexo, 16 May, Maracaibo, Edo. Zulia, Venezuela. Available at: http://rafaelramirez.desarrollo.org.ve/articulos/pdvsafielallegadodeChávez/ (accessed: 12 August 2015).

Ramírez Carreño, Rafael Darío (2013) *Discurso del Ministro del Poder Popular de Petróleo y Minería y presidente de PDVSA, Rafael Ramírez, durante la 163ª Reunión de la Conferencia Ministerial de la OPEP*, 31 May, Caracas. Available at: http://rafael ramirez.desarrollo.org.ve/?s=Discurso+del+Ministro+del+Poder+Popular+de+Petr%C 3%B3leo+y+Miner%C3%ADa+y+presidente+de+PDVSA%2C+Rafael+Ram%C3%A Drez%2C+durante+la+163a+Reuni%C3%B3n+de+la+Conferencia+Ministerial+de+la +OPEP&x=0&y=0 (accessed: 12 August 2015).

Ramírez Carreño, Rafael Darío (2013) *Memoria y Cuenta de los Ministros de la Vicepresidencia del Área de Desarrollo Territorial*, 4 November, Caracas. Available at: http://rafaelramirez.desarrollo.org.ve/articulos/comparecenciadeministrosparael desarrolloterritorial/ (accessed: 12 August 2015).

Ramírez Carreño, Rafael Darío (2014) *Sesión Especial Con Motivo De La Conmemoración Del Centenario Del Inicio De La Actividad Comercial Petrolera En Venezuela, Con La Explotación Del Pozo Zumaque I*, 5 August, Maracaibo. Available at: www. asambleanacional.gob.ve/uploads/documentos/doc_51637a01209475f34f71c1e204a24 cf61e97c0fc.pdf (accessed: 12 August 2015).

Sansó de Ramírez, Beatrice (2007) *Abra Solar de Alejandro Otero*, 9 November, Caracas. Available at: www.pdvsalaestancia.com/?p=1549 (accessed on 14 August 2015).

Sansó de Ramírez, Beatrice (2008) *Develación de la pieza escultórica Los Cerritos*, 12 April, Caracas. Available at: www.pdvsalaestancia.com/?p=1562 (accessed: 12 August 2015).

Sansó de Ramírez, Beatrice (2008) *Discurso con motivo de la develación de la Fisicromía del maestro Carlos Cruz-Diez*, 13 March. Available at: http://pdvsa.com/index. php?tpl=interface.sp/design/biblioteca/readdoc.tpl.html&newsid_obj_ id=6451&newsid_temas=111.= (accessed: 12 August 2015).

Sansó de Ramírez, Beatrice (2008) *Discurso con motivo de la inauguración de la exposición José Félix Ribas, Héroe de Juventudes*, 12 February, Caracas. Available at: www. pdvsalaestancia.com/?p=1560 (accessed: 12 August 2015).

Sansó de Ramírez, Beatrice (2008) *Discurso con motivo de la entrega de instrumentos musicales a niños y niñas de la coral SENECA*, 30 July, Margarita. Available at: www. pdvsalaestancia.com/?p=1581 (accessed: 12 August 2015).

Sansó de Ramírez, Beatrice (2008) *Día de la Mujer*, 8 March, Caracas. Available at: www.pdvsalaestancia.com/?p=1557 (accessed: 14 August 2015).

Sansó de Ramírez, Beatrice (2008) *Discurso con motivo de la entrega de instrumentos musicales a niños y niñas de la coral SENECA*, 30 July, Margarita. Available at: http:// pdvsa.com/index.php?tpl=interface.sp/design/biblioteca/readdoc.tpl.html&newsid_ obj_id=6957&newsid_temas=111. (accessed: 14 August 2015).

Sansó de Ramírez, Beatrice (2008) *Fisicromía del maestro Carlos Cruz-Diez*, 13 March, Caracas. Available at: www.pdvsalaestancia.com/?p=1573 (accessed: 14 August 2015).

Sansó de Ramírez, Beatrice (2008) *Inauguración de PDVSA La Estancia Paraguaná*, 13 August, Paraguaná, Edo. Falcon. Available at: www.pdvsalaestancia.com/?p=1578 (accessed: 14 August 2015).

Sansó de Ramírez, Beatrice (2008) Mateo Manaure el hombre y el artista, 17 July, Caracas. Available at: www.pdvsalaestancia.com/?p=1576 (accessed: 14 August 2015).

Sansó de Ramírez, Beatrice (2008) *Palabras de introducción al concierto de la Camerata Romeu con motivo de la celebración del Día de la Mujer*, 8 March, Caracas. Available at: www.pdvsalaestancia.com/?p=1555 (accessed: 14 August 2015).

Sansó de Ramírez, Beatrice (2008) *Manuel Espinoza Poética del Paisaje*, 6 November, Caracas. Available at: www.pdvsalaestancia.com/?p=1570 (accessed: 14 August 2015).

Sansó de Ramírez, Beatrice (2009) *Inauguración de la Quinta versión de la Fuente de Plaza Venezuela*, 10 March, Caracas. Available at: www.pdvsalaestancia.com/?cat=16&paged=3 (accessed: 14 August 2015).

Sansó de Ramírez, Beatrice (2009) *Inauguración de la exposición Juan Calzadilla Poética visiva y continua*, 9 June. Available at: www.pdvsalaestancia.com/?p=1592 (accessed: 14 August 2015).

Sansó de Ramírez, Beatrice (2009) *Estreno del documental América Tiene Alma del director de cine Carlos Azpúrua*, 10 November. Available at: www.pdvsalaestancia.com/?p=1596 (accessed: 14 August 2015).

Sansó de Ramírez, Beatrice (2010) *Presentación de la Exposición La Pintura en el Espacio del Maestro Omar Carreño*, 3 March, Caracas. Available at: www.pdvsalaestancia.com/?p=1602 (accessed: 14 August 2015).

Sansó de Ramírez, Beatrice (2010) *Inauguración de la sede de PDVSA La Estancia Maracaibo*, 5 February, Maracaibo, Edo. Zulia. Available at: www.pdvsalaestancia.com/?p=1604 (accessed: 14 August 2015).

Sansó de Ramírez, Beatrice (2011) *A un año de labor PDVSA La Estancia Petróleo que cosecha cultura*, 5 February, Caracas. Available at: www.pdvsalaestancia.com/?p=1608#comments (accessed: 14 August 2015).

Sansó de Ramírez, Beatrice (2012) *Palabras como oradora de orden en la Sesión Solemne del Concejo Municipal del Municipio Carirubana, Estado Falcón*, 8 March, Punto Fijo. Available at: www.pdvsalaestancia.com/?p=4684 (accessed: 14 August 2015).

Sansó de Ramírez, Beatrice (2012) *PDVSA LA ESTANCIA: LA UTOPÍA DE LO POSIBLE*, 14 August, Caracas. Available at: www.pdvsalaestancia.com/?p=3302 (accessed: 14 August 2015).

Sansó de Ramírez, Beatrice (2013) *Dos años llevando cultura al pueblo zuliano a través de la Siembra Petrolera*, 4 February, Maracaibo, Edo. Zulia. Available at: www.pdvsalaestancia.com/?p=8449 (accessed: 14 August 2015).

5 Giant oil workers and the expediency of Culture as Renewable Oil

There is a historical narrative in Venezuela that has regarded the arts, and by extension culture, as a resource to be exploited like oil. The most eloquent is a statement made by Venezuelan visual artist and playwright César Rengifo in an interview to daily newspaper El Nacional while working on a mural for the military in 1973:

> We are like oil: a reserve; but in Venezuela we have yet to be put in motion.
> (Rengifo, 1973, p. 12; translation by the author)

In the midst of the oil boom and The Great Venezuela of President Carlos Andrés Pérez, Rengifo asserts his value as an artist by positioning himself as a mineral resource, to suggest that the Petrostate would invest in artists like him, and culture more generally, only if they were akin to a reserve of crude oil. Rengifo's statement illustrates the clout oil carries in defining the relationship between the Petrostate and culture in Venezuela. The discussion developed in this chapter reveals how a particular understanding of culture is privileged by the national oil industry under Petro-Socialism. It does so by examining the advertisement campaign launched by PDVSA La Estancia in 2013 titled *Transformamos el petróleo en un recurso renovable para ti* (We transform oil into a renewable resource for you) through the semiotic lenses of Charles S. Peirce's *Semiosis* and Roland Barthes' *Mythologies*.

The chapter is divided into three parts. The first part provides a review of George Yúdice's proposition of the expediency of Culture as a Resource which enables its use for economic, social and political purposes. The second part provides an overview and context of PDVSA's advertising campaign; it also describes the visual and textual components of the adverts. The third part develops the theoretical underpinning for the semiotic analysis of the visual element of the adverts. It looks at the public art depicted and the giant oil worker using Charles Peirce's *Semiosis*. This is then followed by the semiotic analysis of the verbal text of the campaign using Barthes' Theory of Mythical speech to elucidate the intended meanings behind the slogan 'We transform oil into a renewable resource for you' and thus what notion of culture is mobilised by PDVSA La Estancia. Finally, the analysis of the visual and verbal components draws on George Yúdice to look into the discursive construction of oil as a renewable resource to examine how the adverts construe the notion of Culture as Renewable Oil. The chapter argues that culture then becomes inextricable from land, akin to a mineral deposit, and tightly controlled by the Petrostate. The inclusion of giant oil workers in the adverts, and their interactions with the urban spaces depicted, point to a re-signification of the city as an oil field, in an explicit attempt at naturalising a direct and somewhat mechanistic relationship between oil, culture and the city.

The expediency of culture as a mineral resource

The analysis of the visual and verbal elements of the adverts draws on George Yúdice's expediency of Culture as a Resource to argue that PDVSA La Estancia discursively renders oil and culture as equivalent by construing what this book defines as the notion of Culture as Renewable Oil, as if culture could be extracted and processed like crude oil. George Yúdice's *The Expediency of Culture: Uses of Culture in the Global Era* (2003) explains the utilisation of Culture as a Resource as an instrument to aid social and economic development. Yúdice argues that culture has acquired, to an extent, the same status as a natural resource as a consequence of the process of globalisation, which has accelerated the transformation of all realms of modern life into a resource. Nonetheless, he argues that the use of Culture as a Resource is not a perversion or a reduction of its symbolic dimension. On the contrary, the expediency of Culture as a Resource is a feature of contemporary life, its transformation traced to a performative force, a style of social relations, generated by diverse organised relations between state institutions and society such as schools, universities, mass media, markets and so on (Yúdice, 2003, pp. 47, 60–61). Yúdice characterises performativity 'as an act that "produces which it names"' revealing the power of discourse to produce realities through repetition (pp. 47, 58) and the particular institutional preconditions and processes by which culture and its effects are produced.

While the term expediency refers to the merely political in regards to self-interest, Yúdice's performative understanding of the expediency of culture 'focuses on the strategies implied in any invocation of culture, any invention of tradition, in relation to some purpose or goal' which is what makes it possible to invoke Culture as a Resource 'for determining the value of an action' (Yúdice 2003, pp. 38). The expedience, or convenience, of Culture as a Resource is what allows its use for economic, social and political purposes. For Yúdice (pp. 279), the expediency of Culture as a Resource has become, in practice, the only surviving definition, becoming impossible not to turn to Culture as a Resource as it is congruent to the way we now understand nature, affecting the way culture is viewed and produced. Similarly, this chapter argues that for the state-owned oil company it would be close to impossible not to turn to culture as a mineral resource.

Yúdice's discussion frames this chapter's approach towards the work of PDVSA La Estancia. PDVSA La Estancia has the privilege of being a founder of culture with direct access to the oil rent. It has an organisational and legal autonomy that means that PDVSA La Estancia doesn't need to negotiate with the government or follow any other agenda than its own. Finally, PDVSA La Estancia can even surpass in financial and political power the jurisdictions of the state and other public institutions whose functions overlap with their work over the city. Thus, for PDVSA La Estancia, culture has the same status as a mineral resource.

Profile of the campaign 'we transform oil into a renewable resource for you'

PDVSA La Estancia launched the campaign *Transformamos el petróleo en un recurso renovable para tí* in early 2013, the year of Hugo Chávez' unexpected death, right at the start of what would have been his fourth term in office. Chávez had outlined an ambitious presidential programme with the objective of transforming Venezuela into a world power of oil energy by expanding the extraction of Venezuela's vast reserves of crude oil (Chávez, 2012, pp. 7, 27; Terán Mantovani, 2014, p. 161). Nonetheless, the majority of the public art and public spaces restored by PDVSA La Estancia and depicted in the adverts predate the arrival of Hugo Chávez to the presidency; they are symbols of the modern oil nation, built and erected during the era renamed as the Fourth Republic (1959–1999), vilified by his Bolivarian revolution.

In July 2014, I interviewed the General Manager of PDVSA La Estancia at their main building in Caracas. At the end of the interview, the Department of Public Relations handed me a CD with a set of photographs of every public art and public space that had been restored by the institution up to that moment, such as Alejandro Otero's Abra Solar, Jesús Soto's Esfera Caracas and Sabana Grande Boulevard.

The set of photographs in the CD form the basis of the campaign. Adding PDVSA La Estancia's logo to the photographs would not have sufficed to evidence the institution's direct involvement in restoring these locations to their original state. To evidence their direct role, PDVSA La Estancia devised visual and linguistic strategies, such as the inclusion of a giant oil worker and the verbal text 'we transform oil into a renewable resource for you', to differentiate itself from the institutions that had traditionally received funds from the oil rent to carry out such works, such as municipalities or the National Heritage Institute.

The campaign consists of 23 posters that depict the public spaces and public art that had been restored by PDVSA La Estancia between 2005 and 2012. Each poster features a giant oil worker clad in red gear, portrayed as if caught in the middle of a working day. The adverts were displayed on most of the PDV petrol stations (owned by PDVSA), on PDVSA La Estancia's main headquarters in Caracas, as well as on their Facebook page (PDVSA La Estancia, 2013).

The visual element of the 23 adverts is a photographic image, composed of public art or an architectural structure and the giant oil worker, complemented by the verbal text. The majority of the adverts feature public art; this is a key point showing not just the regeneration aspect but also the intention to create aesthetic and artistic associations with the state-owned oil company. Nineteen out of the twenty-three spaces depicted are located in Caracas, which highlights that PDVSA La Estancia's investment in urban regeneration has taken place predominantly in the capital city. The public art works are identified by a label that contains the name of the artwork, the name of the artist and its location:

> Plaza Venezuela, Santos Michelena, Caracas, Venezuela
> Abra Solar, Alejandro Otero, Caracas, Venezuela
> Esfera Caracas, Jesús Soto, Caracas, Venezuela
> Fisicromía, Carlos Cruz-Diez, Caracas, Venezuela
> Los Cerritos, Alejandro Otero, Mercedes Pardo, Caracas, Venezuela
> Pariata 1957, Omar Carreño, Caracas, Venezuela
> Uracoa, Mateo Manaure, Caracas, Venezuela
> Venezuela Ocho Estrellas, Ender Cepeda, Edo. Zulia, Venezuela

The labels omit the date they were created, the adverts don't differentiate between Esfera Caracas, created in 1982, from the restored structure of the Gazebo of El Calvario, which dates back to the late nineteenth century. Playgrounds and sports grounds are not provided with any specific location; the sports grounds are simply labelled *canchas* (courts) without providing much more detail. Playgrounds are labelled Parques La Alquitrana (Tar Parks), suggesting that such parks have been built by PDVSA La Estancia all over the country. La Alquitrana is a feminised Spanish term for tar; the parks are named after Venezuela's first oil well that spurted crude oil in 1878, located in the Andean region of Táchira (PDVSA La Estancia, no date). Overall, the adverts depict material spaces of leisure and cultural recreation: playgrounds (childhood), sports grounds (youth), public art and parks (families), all made possible by PDVSA La Estancia, by oil.

But these spaces appear miniaturised by the presence of what is, in effect, the focal point of the campaign: the giant oil worker.

Visual semiotics of oil: giant oil workers and the myth of Culture as Renewable Oil

This section describes the two approaches to visual semiotics deployed to analyse the 23 adverts that compose the campaign '*Transformamos el petróleo en un recurso renovable para ti*': Charles S. Peirce's *Semiosis* and Roland Barthes' *Mythologies*.

Semiotics, simply defined, is the study of signs (Margolis and Pauwels, 2011, p. 320). Signs always represent something; they are in place of something else, the object, to create meaning. A sign does not function as a sign until it is taken as sign of that object. For a sign to exist it must be interpreted; the interpretation of signs is what allows us to know the world. While General Semiotics studies 'sign systems, and communicative processes in general' (p. 298), the specific contexts in which signs are used is the focus of Applied Semiotics, of which Visual Semiotics is a branch. Visual Semiotics do not necessarily encompass all non-verbal communication, geometry, writing or any manifestation we could label as 'visually communicated signs'; the topics covered by Visual Semiotics are pictures, drawings, paintings, photographs, films, posters, diagrams, logograms and maps (p. 298).

Words do not depend on images to be understood, but the meaning of an image can change depending on the words that accompany it. However, pictures 'are superior to verbal communication when spatial configurations have to be represented' (Margolis and Pauwels, 2011, p. 300). In visual semiotics 'that which "stands to somebody for something in some respect"' is the image (p. 301). The processing of information also differs between words and pictures, as we are able to process more visual than verbal information; nonetheless, there is a complementary relationship between verbal and visual data in their 'semiotic potential':

> The superiority of pictures as a medium for the representation of the visible and imaginable world is counterbalanced by the superiority of language for representing the invisible world of sounds, smells, tastes, temperature, or logical relations.
>
> (Margolis and Pauwels, 2011, p. 300)

The meaning of an image can be transformed by the verbal comments attached to it, just as the meaning of verbal communication can be changed, enhanced or obscured by an image. Signs represent something; they are in place of something else called the object. There are three correlative elements that must be contemplated: the sign, its referential object and its meaning, which constitute the 'triadic model of the sign' (Margolis and Pauwels, 2011, p. 301) such as the model of Semiosis proposed by Charles S. Peirce.

Charles S. Peirce's *Semiosis* is based on a triadic model, in which the sign is the unity composed by the object (what is represented), the representamen (how it is represented) and the interpretant (how it is interpreted), 'to qualify as sign, all three elements are essential' (Chandler, 2007, p. 29). The process of Semiosis is produced by the interaction between representamen, object and interpretant. The representamen (similar to Saussure's signifier) is considered by Peirce as a semiotic 'first' associated with a semiotic 'second', the object that is being represented by the sign: 'the object of the visual sign is something once seen, experienced, or imagined'; the association between sign and object lead to a semiotic 'third', the interpretant (similar to Saussure's signified) which is the 'mental or behavioural interpretation of the sign' (Margolis and Pauwels, 2011, p. 302). The representamen mediates between the object and the interpretant; it is, in a sense, the lens through which we view the object. The interpretant must not be confused with the interpreter; the interpretant is produced by the relation between the object and the sign. The interpretant of a visual sign is the idea, action, or reaction that is evoked by the sign (2011, p. 302).

This chapter uses Peirce's approach for its contextuality, his semiosis takes into account the context in which signs are produced and interpreted, defining the sign by its effect on the interpretant:

> Reading pictures is a semiotic process (a process of semiosis). Images are signs that do not only have meanings but also create meanings. The meanings they have are related to the objects of the visual signs; the meaning they create to their interpretants.
>
> (Margolis and Pauwels, 2011, p. 312)

The three elements are necessary for Semiosis to take place. The representamen, the object and the interpretant are functional, rather than ontological terms for Peirce; he was interested not with what signs are but with what they do as signs to establish the point of view for interpretation. Peirce does not 'postulate the existence of the object, the object could be fictitious'; in this sense, the distinction between object and interpretant is 'not one between something material and something mental' since the three elements of the visual sign can be 'mental as well as material' (Margolis and Pauwels, 2011, p. 302). It is a matter of sequence in the semiotic process. In sum, the representamen is the form the sign takes, which can be mental or material; the object is that to which the representamen refers; and the interpretant is the sense made of the sign in the process of signification. For the remainder of this chapter, and for ease of understanding, Peirce's representamen will be referred to as 'sign-vehicle'.

Peirce developed a sophisticated sub-classification of signs. This chapter utilises his most fundamental sign triad: icon, index and symbol, a classification based on the relation of the sign-vehicle with the interpretant. A sign is an icon when it 'is similar to its object' (Margolis and Pauwels, 2011, p. 302), it physically resembles or imitate the object; however, an icon does not 'necessarily refer to real objects', if it does, it is an index. A sign is an index when it has an existential relation with the object, when it is affected by the object, 'they are connected with their objects by a natural cause or a spatial or temporal contiguity' (p. 303); indexes always reference their object. Photographs are considered both iconic and indexical signs:

> Passport photos are indexical signs; they serve to identify their owners. In fact, all photos are indices, because one of the characteristics of some indices is that they are connected with their objects by a natural cause or a spatial or temporal contiguity. Photographs, despite their similarity with their objects, are indexical signs for two reasons: first, they are produced by the physical cause of a light ray projection on a film; second, they serve to identify the object which they depict. The indexicality of the photo does not exclude or contradict its iconicity; the latter is included in the former.
>
> (Margolis and Pauwels, 2011, p. 303)

A sign is a symbol when the relationship is arbitrary or strictly conventional, meaning that this relationship has been 'agreed' upon and learned; for example, language in general is symbolic, and so are numbers, national flags or traffic lights. A symbol is not a physical entity, is a concept, a general rule; a symbol is made up of icons and indexes, all indexes have iconic aspects, the icon being the simplest sign unit.

In the context of this analysis PDVSA La Estancia's adverts are considered fully formed signs, dissected to their smallest units of meaning using Peirce's triadic model. The advert, as a sign unit, will be analysed as a composite of the city space, the giant oil worker and the verbal text. The point of departure for the semiotic analysis is the advert as material sign-vehicle. The analysis will focus on what are considered the two main features of the composition in terms of meaning: the photographic image and the verbal text. The photographic image depicts the interaction of two fundamental features: a giant oil worker and the material space of the city. The verbal text communicates the message conveyed (the interpretant) by this visual interaction. The analysis of the visual elements of the PDVSA La Estancia adverts provide the means to determine what it is exactly that is being represented, and what is the effect, or intended interpretant of the campaign.

To further explore the signification of the PDVSA La Estancia adverts, the visual semiotic analysis of the adverts also draws on Barthes' theory of myths. Roland Barthes was a pioneer in the semiotic study of images, with works focused on photography such as *The Photographic Message* (1961) and *The Rhetoric of the Image* (1964). This section focuses on *Mythologies* (1957), an earlier and still influential work that developed his theory of the myth through the analysis of French adverts.

For Barthes, myth is a type of speech, 'a system of communication, that it is a message', it is not a concept or an idea, it is a concrete entity, a form, a 'mode of signification' defined not by its literal sense but by its intention (1993, pp. 109, 124); in his sense, myth is a type of speech 'chosen by history: it cannot possibly evolve from the "nature" of things' (Barthes, 1993, p. 110). For Barthes, every-thing can be a myth as long as 'it is conveyed by a discourse', it is not confined to the written word or oral, all visual representation mediums such as cinema, photography, reporting and advertising can serve as a vehicle for mythical speech:

> we are no longer dealing here with a theoretical mode of representation: we are dealing with *this* particular image, which is given for *this* particular signification. Mythical speech is made of a material which has *already* been worked on so as to make it suitable for communication: it is because all the materials of myth (whether pictorial or written) presuppose a signifying consciousness, that one can reason about them whilst discounting their substance.
>
> (Barthes, 1993, p. 110)

Barthes proposed myth as a semiological system, as 'one fragment of this vast science of signs which Saussure postulated some 40 years ago under the name *semiology*' (Barthes, 1993, p. 111). Mythology is part of semiology, as a formal science, and of ideology, as an historical science. Mythology derives from Saussure's signifier (carrier of meaning), signified (mental concept of the meaning) and sign (associative total of signifier and signified); for Barthes there are functional implications between the three, there is no arbitrariness, on the plane of myth the signifier is the form, the signified is the concept and the sign is the signification (Barthes, 1993, p. 114). The signifier 'already postulates a reading', a signification is already built and the meaning complete, when the signifier becomes form 'the meaning leaves its contingency behind; it empties itself, it becomes impoverished, history evaporates, only the letter remains' (p. 117). Mythical signification is always motivated; it is never arbitrary as 'there is no myth without motivated form' (p. 126). Signification always contains some analogy, it 'plays on the analogy between meaning and form', an analogy that is supplied to the form by history (p. 126).

The concept is not an abstract element, through it 'a new history is implanted in the myth' and the mythical concept is supposed to be appropriated (Barthes, 1993, p. 119). The concept can take a myriad of forms, making the concept poorer than Saussure's signifier as it 'does nothing but re-present itself' (p. 120) but this multiplication is what allows the myth to be deciphered: 'it is the insistence of a kind of behavior which reveals its intention' (p. 120). The signification or the association of the form and the concept is the myth itself: '*myth hides nothing*: its function is to distort, not to make disappear' (p. 121). Furthermore, the presence of the form is spatial, 'the elements of the form are related as to place and proximity', while the concept appears in a more abstract manner as 'hazy' knowledge it is united to the myth by a relation of deformation as in myth the concept distorts the meaning by, for example, depriving it from its history (p. 122).

Myth does not hide, lie, confess or flaunt: 'its function is to distort, not to make disappear' (Barthes, 1993, p. 121). Myth is an inflexion of the concept, and through this inflexion rather than revealing or dissolving it, naturalises it: 'we reach the very principle of myth: it transforms history into nature' (p. 129). Therefore, myth is read as a reason, never as a motive; it aims to cause an immediate impression, 'experienced as innocent speech not because its intentions are hidden – if they were hidden they would not be efficacious – but because they are naturalized' as factual (pp. 130–131). This factualness characterises myth as 'depoliticized speech' (political understood in its deeper meaning as the power humans have in making their own world); by transforming history into nature contingency appears eternal as things lose their historical quality and memory: 'things lose the memory that they once were made' (p. 142). This chapter posits that PDVSA La Estancia propagates the myth of 'Culture as Renewable Oil', communicated through two modes of discourse: the visual component and the verbal text. Each one of PDVSA La Estancia's adverts is a fully formed composition, fusing image and text. The verbal text fuses the visual interaction between giant oil worker and the city and enriches the meaning.

The context in which myth flourishes is that of the capitalist bourgeois society (Barthes, 1993, pp. 137–138), for in bourgeois culture according to Barthes 'there is neither proletarian culture nor proletarian morality, there is no proletarian art; ideologically, all that is not bourgeois is obliged to borrow from the bourgeoisie' (p. 139). Therefore revolutionary language is not mythical because by definition, revolution excludes myth for it is meant to reveal 'the political load of the world', the revolution aims to make the world and be absorbed in the making of the world; it is political in all its senses, unlike myth which 'is initially political and finally natural' (pp. 143–146).

In sum, Peirce's Semiosis is used to focus on the analysis of the visual component of the advert and the verbal text and Barthes's Myth provides the analytical framework to characterise the language of the verbal text of the adverts as mythical speech. The text is dissected to argue that the adverts' construction of mythical speech is abolishing the history of the public art and public spaces depicted while aiming to naturalise the giant oil workers as agents of what is in essence a paradoxical proposition: the transformation of oil into a 'renewable resource'.

Giant oil workers and the myth of Culture as Renewable Oil

The set of professional photographs provided by the Department of Public Relations of PDVSA La Estancia are part of a larger set intended to be used as stock for promotional material. Most of the spaces were photographed at dusk, under dark blue skies, the night lights blurring the hustle and bustle of motorised traffic and people at rush hour. As these are all pre-existing structures, the photographs by themselves would not evidence PDVSA La Estancia's direct involvement in restoring the depicted public spaces and public art to their original state, after many years of disrepair and vandalism. With a few exceptions, the public art and public spaces restored by PDVSA La Estancia predate Hugo Chávez's presidencies. PDVSA La Estancia alters the photographs of actual, identifiable material spaces of the city, by including the giant oil worker through digital manipulation. This visual fusion engages in a direct interaction with the public space/public art depicted. However, the main focal point of the adverts is not the restored structures but the giant oil worker, who signals the viewer to look at what she/he is working on: the public art and public spaces restored by PDVSA La Estancia.

Each one of PDVSA La Estancia's adverts is a fully formed sign-vehicle in Peirce's terms. The two main features of the composition in terms of meaning are the photographic image and the verbal text. The photographic image depicts the interaction of two fundamental elements: a giant oil worker and the material space of the city. The verbal text communicates the message (the interpretant) of the visual interaction between the giant oil worker and the city. The photographs used in the adverts are indexical because they have a similarity to their objects, the actual spaces of the city intervened by PDVSA La Estancia, easily identifiable. Their indexicality is emphasised by the label attached that identifies each piece of public art or location. Beyond the indexicality of the photograph, the density of meaning of the advert is brought by the inclusion of the giant oil worker through digital manipulation, who engages in a direct interaction with the public space/public art depicted.

The giant oil worker is both an iconic and indexical sign. Iconic for its similarity to an actual oil worker (a human-scale one, as actual giant oil workers do not exist) and indexical because of their relationship with the oil industry. The giant oil worker suggests the totality of the oil industry, mainly the extraction of crude oil. The oil workers are the main feature of the sign-vehicle, they appear in full working gear, clad in red from head to toe (red is the colour that identifies the Bolivarian revolution), with hard hats, protective overalls, boots, gloves and wielding tools. Their actions suggest the undertaking of heavy duty work, they are wielding tools such as pipe wrenches and protective gloves, directly manipulating the public art or architectural structures as if they were heavy machinery in a refinery, the actual natural environment of an oil worker. Although they are of an unnatural monumental size, collectively the oil workers are portrayed as benevolent giants occupied with beautifying the city, presented as a naturalised and ubiquitous presence in a city miniaturised by their scale. All the adverts vividly capture the face of each giant oil worker, their portraits are predominantly three-quarter or full body photographs, their monumentality emphasised by the low-angle shot; they do not gaze directly at the viewer, they are either focused on their duty or interacting with miniaturised city inhabitants below them in a cordial manner.

In the pair of posters of Sabana Grande Boulevard, the giant oil workers are seen installing a canopy. In a space that has been aptly named Plaza Siembra Petrolera (Sowing Oil Square), they are presented as if they were turning the giant pipe of an oil drill into the ground. In the case of all adverts that depict public art, they lose their monumental scale through the interaction with the giant oil worker. These spaces, all completed, are shown as a work in progress, as if the giant oil workers were caught in the midst of repairing them.

As Iconic-Indexical signs, the giant oil workers also function as visual-cultural ambassadors of PDVSA La Estancia. Instead of the *ejecutivo petrolero* (oil executive), the Petro-Socialist and revolutionary PDVSA chose to be represented by the *obrero petrolero* (oil worker), the one who gets down and dirty to perform the extraction of crude oil from the subsoil. Overall, this strategy is a continuation of the legacy left by the foreign oil companies, such as Shell and Creole, who inaugurated the use of Public Relations in Venezuela:

> The departments of public relations of Creole and Shell in Venezuela had multiple functions: internally they promoted practices considered useful among foreign and Venezuelan employees, whilst externally they served as the public face of the company. These tasks were not mutually exclusive because these companies viewed employees as their ambassadors.
>
> (Tinker Salas, 2009, p. 281, translation by the author)

If the giant oil workers are the visual-cultural ambassadors of PDVSA, then their inclusion and interaction with the spaces depicted point to a re-possession of the city, visually reconstructed as an oil field in a clear attempt at naturalising a direct and mechanistic relationship between oil, culture and the city. The giant oil workers also function as a visual metaphor of PDVSA's State Space, transferring the extractive activity of the oil field to the city, mediated by public space and public art. As an artificially constructed image, the giant oil worker builds a direct connection between the oil company and the viewer, and within the advert between the giant oil worker and city dwellers, suggesting a symbolic break with the historical social division between workers of the oil industry and the rest of the population:

> Many directly employed or indirectly benefiting from the oil industry subsequently assumed key positions in Venezuelan society, government, commerce and industry. Their views reflected a series of self-sustaining myths about the oil industry and its importance to the nation and society. Paramount among these was the notion that for Venezuela the oil industry was the means to achieve modernity in all its forms. For those employed by the industry, these new modern traditions accentuated certain traits and behaviour patterns – discipline, efficiency, work ethic, meritocracy, and in some cases even bilingualism – that helped define the 'collective consciousness' of the oil industry and distinguished those working in it from the rest of society.
>
> (Tinker Salas, 2009, p. 5)

But this symbolic break with the social hierarchy that used to distinguish those working for the oil industry with the rest of society is contradicted by the colossal scale of the giant oil workers and their demeanour towards the viewer. The giant oil workers do not address or acknowledge the viewer, and their gaze does not demand anything from the viewer (Kress and van Leeuwen, 2006, pp. 121–122); rather, their scale incites deference. Furthermore, the campaign romanticises the harsh reality and dirtiness of work in the oil fields by offering a sanitised version of the oil workers as they are relocated in the midst of the city. The giant oil worker appears then as a symbol of PDVSA's Bureaucratic Power and benevolence, they represent oil, once again, as the source of prosperity and the producer of culture through the labour of a 'workers elite'. The signification in the adverts is that PDVSA is the sole provider of prosperity and culture, materialised in the beautiful restored public spaces and public art.

The relocation of the sanitised version of the giant oil worker turns the adverts into symbolic-signs (defined by Peirce as a relationship that has been 'agreed' upon and learned) of the Oil Social District as PDVSA's State Space, a symbolic-sign of the city colonised by oil and appropriated by PDVSA. Furthermore, seen together, the 23 adverts represent a cause and effect relation (Margolis and Pauwels, 2011, p. 308) of the impact of the work of PDVSA La Estancia on the city. Two types of cause and effect relations can be identified in the adverts: a natural one and an artificial one. The natural is that those spaces were in fact restored by PDVSA La Estancia by its self-defined revolutionary oil workers, all directly funded by the oil rent. But the giant oil worker in the advert also serves to obscure the bureaucratic mechanisms and institutional structures needed for restorations of this scale to come to fruition, but only the authorship of PDVSA is acknowledged. Visually, it is not as simple as having a giant oil worker do the repairs, public works need sophisticated and complex sets of decisions and institutional arrangements to make them possible. However, there is some truth to this portrayed simplicity; the mechanisms that led to the progressive dismantlement of the existing institutional apparatus, discussed in Chapter 3, opened a breach that allowed PDVSA La Estancia to claim ownership over public spaces and public art, without having to acknowledge the set of institutions and bureaucratic logistics that had to come together to make these works possible in the first place. The artificial cause and effect lies in claiming that they are transforming oil into a 'renewable resource'. The term 'renewable resource' has been used interchangeably by PDVSA to refer either to oil or culture. The magazine *Petróleo y Revolución #14* (Oil and Revolution, May 2012), an internal publication produced and distributed by the Ministry of Popular Power of Oil and Mining, includes an article that reviews the work of PDVSA La Estancia as 'making our culture a renewable resource'.

According to Peirce, images 'cannot express the ideas of possibility, necessity, obligation, or volition' (Margolis and Pauwels, 2011, p. 309); while the visual elements of the adverts cannot tell us the motivations or subsequent effects of the giant oil workers' engagements with the material space of the city, the verbal text aims to do just this, it provides the images with volition, it expresses the idea of possibility latent in the advert. It is the visualisation of the 'utopia of the possible' discussed in Chapter 4, a 'realised' utopia of inexhaustible oil dreamt by Chávez, obtained by harvesting culture from sowing oil, constructed as an illusion of the New Magical State.

The verbal text that identifies the campaign suggests the correlation between the myth of 'renewable oil' put forward by the adverts and the instrumentalisation of culture. This chapter now turns to develop a semiotic analysis of the verbal text to identify the 'mythical speech' of the adverts, using Roland Barthes' *Mythologies*. Barthes provides the analytical framework to decipher the type of reality that the language of the adverts aims to create, and hence what notion of culture is construed by PDVSA La Estancia.

The slogan '*Transformamos el petróleo en un recurso renovable para ti*' (We transform oil into a renewable source for you), expresses eloquently the purpose of the campaign, it reveals the verbal text as the construction of myth in Barthes' terms. The myth is completed by the interaction between the verbal text and the visuals of the campaign. The campaign presents as fact that PDVSA La Estancia is transforming oil into a 'renewable resource' but it does not make explicit what this renewable resource is, rather it is suggested in the images of the oil workers 'sowing oil' in the city. The intention of the campaign was already suggested by PDVSA La Estancia's discursive construction that claims it is 'oil that harvests culture' in the city.

The myth is emphasised in the colouring and layout of the typography. The words of the verbal text are given two different colour combinations, white and red, or black and red. The same three words are always highlighted in red and bold in every single advert:

> We transform ***OIL***
> into a resource **RENEWABLE** FOR **YOU**

The implied '*nosotros*' (we) in the plural of *Transformamos* (we transform) in the verbal text is the voice of the institution represented by the collective voice of the giant oil workers depicted in the adverts, the visual-cultural ambassadors of PDVSA La Estancia. The 'We' also excludes viewers. Compounded by the monumental scale of the giant oil workers, the use of 'we' hales viewers as the beneficiaries of PDVSA's philanthropy, setting them as one-down to the company. Then, *Petróleo* (oil) is not only highlighted, but appears in a larger size, singled out visually. In the typographic layout, the left half of the text is written in lowercase letters while the right half dominates the composition with words in capital letters and highlighted in bold red: ***OIL RENEWABLE FOR YOU***. The key message of the campaign is contained in the right half. The composition of the verbal text produces a focus of information; Kress and van Leeuwen propose that as a general rule in a visual composition, the elements placed on the left are presented as Given, and the elements placed on the right are presented as New:

> For something to be Given means that it is presented as something the viewer already knows, as a familiar and agreed-upon point of departure for the message. For something to be New means that it is presented as something which is not yet known, or perhaps not yet agreed upon by the viewer, hence as something to which the viewer must pay special attention.
>
> (Kress and van Leeuwen, 1996, p. 187)

The position of *Petróleo* (oil) and *renovable* (renewable) in the layout of the text puts *Petróleo* right on top of *renovable*, which visually suggests them to be read together as *petróleo renovable* (renewable oil). The meaning of oil and renewable put together is paradoxical because oil is a non-renewable natural resource, it is finite. But the paradox implied here is not literal. While the paired words suggest that PDVSA La Estancia is claiming that a never-ending supply of oil is made possible as PDVSA succeeds in 'magically' transforming oil into a renewable resource, oil is made renewable metaphorically by investing the oil rent in culture. Hence, if oil is made renewable by investing in culture then culture is like renewable oil: oil, culture and the city are conflated into one entity, thus disputing Rodolfo Quintero's assertion that an 'oil city' does not produce art, science or any form of intellectual culture when the inert and money-driven 'culture of oil' prevails (Quintero, 2011, p. 55) by coalescing oil, culture and the material space of the city, making an expedient use of culture in Yúdice's terms.

The next words highlighted in bold red lead to the intended audience of the campaign, *FOR YOU*. It's made renewable *for you*. The 'for you' aims to create identification, a sense of closeness between the institution and the viewer by addressing the viewer directly. The audience to which 'YOU' is addressed to is the potential users of the spaces, the dwellers of the cities these structures are located in, but given the national scope of the campaign, it implies the population of Venezuela at large, in other words, all citizens of the Petrostate who share oil as a national identity and are the subjects of the New Magical State.

The adverts as a set are the material form of the sign, and it is this multiplication that allows us to decipher the myth through the repetition, in this case, of the idea of transforming oil into a 'renewable resource'. The concept relies on understanding culture as crude oil, as an expedient mineral resource that accumulates in the subsoil that upon extraction yields political and social benefits, but that unlike oil can be transformed and renewed. Oil then becomes a double-sided cultural entity; it is invoked as a singular force capable of producing palpable cultural effects: a producer of culture and a cultural product in itself. If culture is like a mineral resource extracted from the subsoil of the nation, then culture and the discursive and political wealth produced by Culture as Renewable Oil is, like the subsoil, the exclusive property of the Petrostate.

The myth presented by the adverts is oil made renewable by investing in culture, thus culture becomes renewable oil. The myth comes into fruition through the performative force of PDVSA's giant oil workers. The myth of 'renewable oil' is founded on the certainty that Venezuela possesses the world's largest crude oil reserves, certified by OPEC in 2011. In the manner of the demiurgic exercise of power of the Magical State defined by Coronil (1997), PDVSA La Estancia becomes the sorcerer that makes possible the impossible by transforming oil into a renewable resource. Oil and culture are rendered as equivalent when culture is turned into crude oil, as if culture was a physical material that could be industrially extracted, processed and distributed. Moreover, 'renewable oil' attaches to culture conflicting notions of resource: resource as nature and resource as deposits extracted from the subsoil that yield wealth.

In this context, the notion of Culture as Renewable Oil is more than a discursive construction. The implication is that by tying culture to the soil, it constructs culture as a material entity that accumulates in the subsoil, ready to be extracted, processed, weighed and measured under tight control by the Petrostate. Therefore, modern and abstract manifestations of culture such as the ones depicted in the adverts can only be quantified as valuable if they are transformed into oil machinery, the instruments of the country's modernisation and progress. Then, for culture to fit this rhetoric, it has to be defined in narrower terms and become the exclusive property of the Petrostate. To paraphrase Yúdice, it is impossible for the Venezuelan state-owned oil company not to view culture as a reserve of crude oil. While Yúdice (2003) looks at performativity in grassroots movements of Latin America as a resistance to neoliberal strategies, in this case it is the 'anti-neoliberal' giant oil workers of Petro-Socialism that are being called to perform culture. In the Petro-Socialist notion of Culture as Renewable Oil, culture is bound to the subsoil, so that it becomes the exclusive property of the Petrostate and belongs to the realm of PDVSA's State Space.

PDVSA La Estancia's instrumentalisation of Culture as Renewable Oil is not oriented towards the conventional model of development discussed by Yúdice, rather it is instrumentalised for political aims. Under Petro-Socialism culture is not required to provide financial returns, as the Petrostate's distributive expenditure policies do not require financial accountability or representative bargains. PDVSA La Estancia conceives culture as a medium to transcend the ills of capitalist urban living. Culture as Renewable Oil condenses the materiality lost by oil when it was reduced to the fetish of rent money (Pérez Schael, 1993, p. 94) becoming inextricable from land. The invocation of Culture as Renewable Oil becomes the fuel to extend the life of the illusions of the New Magical State beyond the physical death of Chávez.

Conclusion

The campaign is a symbol of PDVSA's discursive power. The space of the city is envisioned as an oil field, where giant oil workers are seen repairing or installing architectural or sculptural structures as if they were pieces of heavy machinery in an oil refinery. They give a face –or rather a collective face – to oil in the campaign as their colossal figures subliminally communicate that the urban spaces, and the city at large, are contained within the realm of PDVSA.

The campaign can also be interpreted as a sign, through the giant oil worker, of the Oil Social District, PDVSA's State Space. PDVSA La Estancia is a national management entity of PDVSA; its jurisdiction and cultural policy founded on the Organic Law of Hydrocarbons. Caracas, as discussed in Chapter 3, is covered by the Metropolitan Oil District, spanning across the Metropolitan District. Caracas is not, and has never been, a site of oil extraction and refining; there are no crude oil deposits in its subsoil but it is the site of PDVSA's headquarters and the seat of national government. Hence, if PDVSA's headquarters are considered a site of 'oil extraction', then the Metropolitan Social Oil District encloses Caracas under its sphere of influence, which underpins PDVSA La Estancia's interventions depicted in the adverts. The material space of the city is envisioned as an oil field, where giant oil workers are seen repairing or installing architectural or sculptural structures. The giant oil workers provide a collective human face to the national oil industry as their colossal figures subliminally communicate that urban spaces, and the country at large, are contained within PDVSA's State Space.

Furthermore, the spaces in the adverts are, in Peircean terms, tokens of the cultural effects of oil that could potentially expand to other regions covered by the Oil Social Districts. PDVSA La Estancia devised the visual strategy of the giant oil worker to establish their ownership over urban spaces. The naturalisation of the giant oil workers metaphorically harvesting culture, acting outside of its natural environment, the oil fields, and stripped of the dirtiness of directly handling oil makes them appear more analogous to gardeners than actual oil workers. The adverts visualise the performance of 'sowing' oil and 'harvesting' culture, using the giant oil worker as the personification of Bauman's 'manager-farmer'. This performativity is also made possible by a State Space in flux, in transition towards the Socialist State; hence, if culture is akin to a mineral resource it becomes inextricable from State Space. The labour of the giant oil workers transform oil into a 'renewable resource' as the new illusion of the New Magical State, congruent with PDVSA La Estancia's 'utopia of the possible' using culture expediently to conceptualise it as Culture as Renewable Oil. This evokes a farming cycle that on one hand, responds to the discursive strand of 'renewable oil' and, on the other, provides a new illusion, a novel dramaturgical act for the New Magical State: Culture as Renewable Oil is tied back to the land. Hence, Culture as Renewable Oil becomes inextricable from State Space, from territory, specifically from the Oil Social District as PDVSA's parallel State Space. If culture is 'harvested' from the subsoil, then the Petrostate can claim complete ownership and tight control over culture as a 'renewable resource' as established by the Organic Law of Hydrocarbons.

This is the context of the naturalisation of the giant oil worker metaphorically transforming oil into culture, acting outside of its natural environment – the refinery and oil fields – and stripped of the dirtiness of handling crude oil. Their colossal scale provides them with a mythical aura, while at the same time performing the material quality that oil had lost by being reduced to rent money. The notion of Culture as Renewable Oil is bound to the territory of the world's largest crude oil reserves, essential for the survival of Petro-Socialism. In the specific context of this campaign, the verbal text suggests that oil is being transformed into culture, inferring that culture is the quintessential renewable resource, naturalising as truth that PDVSA La Estancia is making possible the utopia of inexhaustible oil.

References

Barthes, R. (1993) *Mythologies*. London: Vintage.

Chandler, D. (2007) *Semiotics: The Basics*. Routledge.

Chávez, H. (2012) *Propuesta del Candidato de la Patria Comandante Hugo Chávez para la gestión Bolivariana socialista 2013–2019*, *www.mppeuct.gob.ve*. Caracas, Venezuela: Comando Campaña Carabobo. Available at: www.mppeuct.gob.ve/el-ministerio/politicas/leyes-y-planes/propuesta-del-candidato-de-la-patria-comandante-hugo-chávez (accessed: 24 March 2015).

Coronil, F. (1997) *The Magical State: Nature, Money, and Modernity in Venezuela*. Chicago: University of Chicago Press.

Kress, G. and van Leeuwen, T. (1996) *Reading Images: The Grammar of Visual Design*. New York: Routledge.

Kress, G. and van Leeuwen, T. (2006) *Reading Images: The Grammar of Visual Design*. New York: Routledge.

Margolis, E. and Pauwels, L. (2011) '16 Visual Semiotics: Key Features and an Application to Picture Ads', in Margolis, E. and Pauwels, L. (eds) *The SAGE Handbook of Visual Research Methods*. Los Angeles, CA: Sage Publications Ltd, pp. 298–317.

PDVSA La Estancia (2013) *Facebook, TRANSFORMAMOS EL PETRÓLEO EN UN RECURSO RENOVABLE PARA TI*. Available at: www.facebook.com/media/set/?set=a.296539377135848.64316.130175993772188&type=3 (accessed: 5 September 2014).

PDVSA La Estancia (no date) *Eje Patrimonial.* Available at: www.pdvsalaestancia. com/?page_id=462 (accessed: 14 April 2016).

Pérez Schael, M. S. (1993) *Petróleo, cultura y poder en Venezuela.* Caracas: El Nacional.

Quintero, R. (2011) 'La cultura del petróleo: ensayo sobre estilos de vida de brupos sociales de Venezuela', *Revista BCV*, pp. 15–81.

Rengifo, C. (1973) 'Los artistas en Venezuela somos como el petróleo: una reserva', *El Nacional*, sección C, página 12.

Terán Mantovani, E. (2014) *El fantasma de la Gran Venezuela: un estudio del mito del desarrollo y los dilemas del petro-Estado en la Revolución Bolivariana, CLACSO.* Caracas, Venezuela: Fundación CELARG. Available at: www.clacso.org.ar/libreria-latinoAmericana/libro_detalle.php?id_libro=1092&pageNum_rs_libros=0&totalRows_rs_libros=1057 (accessed: 22 March 2016).

Tinker Salas, M. (2009) *The Enduring Legacy: Oil, Culture, and Society in Venezuela.* Durham: Duke University Press.

Yúdice, G. (2003) *The Expediency of Culture: Uses of Culture in the Global Era.* London: Duke University Press.

Conclusion

The untenable utopia of oil

For a nation whose identity is bound with oil and with an economy entirely dependent on oil, even in the midst of the current political, economic and social debacle, oil remains the only plausible path towards prosperity. The memory of the glories of the Magical State lurk behind the legacies of the spectacular modernism of the 1950s, the rapid urbanisation and modernisation in the 1960s, the oil boom of the 1970s showcased in the most enviable urban and cultural infrastructure and architecture in the region, brought to a traumatic halt with the steady decline in oil prices of the 1980s and 1990s. The coincidental unprecedented rise in oil prices and Hugo Chávez's presidencies marked another pivotal shift in the oil boom and bust history of the Venezuelan Petrostate. The oil windfall allowed Chávez to embark on an ambitious plan to transform the country into a Petro-Socialist State, with the state-owned oil company PDVSA playing a fundamental role in building his alternative model of development. The Petrostate was at the centre of Chávez political project, for whom oil rentierism became a vital condition to lead Venezuela towards a new socialist society. Chávez reinstated the Petrostate at the centre of his political project, presenting oil rentierism as the essential condition to lead Venezuela towards an alternative model of development. He turned PDVSA into the fundamental instrument of revolutionary change, handing over many government functions and new social programs to the state-owned oil company (ranging from food distribution and social housing, to city spaces and cultural objects), in the process eroding the bureaucratic structure consolidated during the democratic period (1958–1998). Chávez's ambition to transform Venezuela into a world energy power (Terán Mantovani, 2015, p. 112) stood on the belief that the oil windfall would be everlasting. While the Fourth Republic trailed the path to 'sow the oil' Chávez's Bolivarian revolution would achieve what previous governments could not: the 'harvest' of oil. He spoke from the standpoint of the new sole owner of the 'seed' (oil) vindicating oil revenue for collective benefit but his 'harvest' of oil was an act of appropriating already existing oil wealth, not the production of new sources of wealth (Urbaneja, 2013, pp. 81–89). His New Magical State stood on an ephemeral foundation on which to build a new kind of socialist state which would translate, in the end, in the bankruptcy of Venezuela's modernisation project.

Although Chávez championed an anti-capitalist and anti-neoliberal agenda, the shift in the relationship between the state and the state-owned oil company brought by Petro-Socialism blurred the boundaries between the state and PDVSA. Once PDVSA was reframed as a 'revolutionary' oil company at the service of Petro-Socialism, Rafael Ramírez, former President of PDVSA, publicly confirmed that the oil company was at the service of the construction of the Socialist State and the eradication of oil rentierism. It is contradictory for the president of an oil company to publicly condemn oil capitalism and argue for the end of oil rentierism. Nonetheless, Ramírez established PDVSA's public identity as a revolutionary corporation, even though in practice it had to preserve its active engagement in global oil markets. Furthermore, Petro-Socialism was heavily dependent on the longevity of OPEC and a continued growth of global oil capitalism. The entrenchment of oil rentierism was necessary to guarantee the irreversibility of socialism; Chávez's policies were designed on the premise of inexhaustible highly priced oil.

An essential contradiction runs through Petro-Socialism: the alternative model to capitalism was in practice heavily reliant on the success of global oil capitalism, the very model Chávez proposed to eradicate. But in terms of his exercise of Bureaucratic Power it gained some coherence given that only an oil rentier state could produce a New Magical State; its powers and longevity bound to a 'revolutionary' PDVSA that provided the necessary resources to consolidate the Socialist State. The oil windfall and the reforms of the state apparatus enabled Hugo Chávez to summon all the bureaucratic powers of the state to embody a New Magical State in his persona to transform the country into a socialist society (Coronil, 2011, p. 9). Despite Chávez's anti-capitalist and anti-neoliberal rhetoric, Venezuela could never cease to be an oil rentier state.

Hugo Chávez formally launched Petro-Socialism at the beginning of his third presidential term (2007–2013), presented as an economic and political model that uses the vast financial resources generated by the oil windfall to build a socialist state and a new socialist society. His socialist state needed to consolidate its own Socialist State Space (Brenner and Elden, 2009), crucial not only in terms of the deployment of Bureaucratic Power but most importantly to secure control and authority over the subsoil from where oil wealth and the powers of the New Magical State originated.

The Constitution of the Bolivarian Republic of Venezuela, the development plans for the nation and the new laws manifested Chávez's will to create a new Socialist State Space through the principles of the New Geometry of Power. To this end he proceeded to dismantle existing bureaucratic structures that were deemed an unnecessary obstacle. Although the New Geometry of Power was proposed as the guiding principle for the creation of the spatial policies of the socialist state, in practice, Chávez government proved to be inefficient in clearly delineating a cohesive conceptual framework to constitute a socialist State Space. In lieu of a cohesive territorial strategy, PDVSA was able to exercise and impose the Oil Social District outlined by the Organic Law of Hydrocarbons as a parallel State Space, extending its dominant space beyond the spatial enclaves of the oil industry.

The deficiency in the development of coherent spatial policies, such as the specific case of the city of Caracas, were compounded by the chequered transition towards the socialist State Space. The fast pace of abrogation, creation and non-enactment of the reform of the laws of territorial planning of the 1980s, as well as the creation of new laws for the Socialist State that contravened the Constitution of the Bolivarian Republic of Venezuela engendered a chaotic institutional landscape. The discontinuous process of dismantlement of the state apparatus opened a breach in the legal framework of territory between the abrogation of the legal instruments of territorial planning in 2005, and the creation of the legal framework for the Communal State in 2010, which enabled PDVSA La Estancia, the social and cultural arm of PDVSA, to override the fragmented institutional landscape by abiding to the Organic Law of Hydrocarbons, particularly to Article 5 and the Oil Social Districts. The vacuum left by the transition towards the socialist State Space provided the ideal circumstances to conceptualise and enact the Oil Social District as PDVSA's State Space. Since PDVSA had been effectively functioning as a parallel state, this confirms Brenner and Elden's contention that there is no state without territory and no territory without a state. PDVSA's parallel State Space raises broader issues on the state-owned oil company's ownership and authority over the material space of the city. As the producer of the oil rent PDVSA controls the main source of income of the state; its vast financial resources and managerial efficiency both challenged and diminished the authority of municipalities and regional governments that still abided to the abrogated laws of territorial organisation and urban planning of 1983 and 1987, respectively. The fault lines created by the conflicting coexistence of the three models of State Space (the Constitution of the Bolivarian Republic of Venezuela, PDVSA and Communal State) enabled PDVSA to extend the dominion of the Metropolitan Oil Social District across Caracas, consolidated by the interventions of PDVSA La Estancia, the social and cultural arm of PDVSA.

The case of Caracas, whose fragmented legal and institutional framework of urban governance carried deficiencies inherited from previous governments, serves to illustrate the discontinuities of the transition and the mechanisms that ultimately enabled PDVSA La Estancia to use the Organic Law of Hydrocarbons to supersede the legal authority of municipalities, and implement it as an implicit cultural policy. Caracas is not, and has never been, a site of oil extraction and refining. Nonetheless, the Metropolitan Oil District encloses Caracas because that is where the headquarters of PDVSA is located, absorbing spaces such as the Sabana Grande Boulevard and Plaza Venezuela into the areas of influence of PDVSA's State Space. It is a symptom of the vacuum left by the dismantlement and abrogation of the policy instruments and institutions inherited from the Fourth Republic and the slow and discontinuous transition towards the Socialist State. The discontinuous process of abrogation and creation of spatial policies generated fault lines in the bureaucratic apparatus of urban governance in Caracas, as well as a chasm between the Absolute-Representations of Space sanctioned by the Constitution of the Bolivarian Republic of Venezuela and the new policy instruments of the Socialist State.

Given this chasm, it is primordial to revisit Rodolfo Quintero's definition of the 'oil city' (Quintero, 2011, pp. 46–55) as a institutionally deprived and culturally sterile city in relation to the literature on rentier states that argue that oil rentierism, particularly during cycles of oil booms, hinders institutions, alters fiscal and bureaucratic structures and diminishes the capacity of the state (Beblawi and Luciani, 1987; Karl, 1997; Maass, 2009; Corrales and Penfold, 2011; Mitchell, 2011; Ross, 2012), which contradicts the historical attributes of oil as a carrier of progress and modernity in Venezuela (Pérez Schael, 1993; Coronil, 1997; Tinker Salas, 2009).

PDVSA La Estancia articulated the ambition to transform Caracas into a 'socialist city, paradigm of the twenty first century'. By initially defining Caracas as an oil city in Quintero's terms, PDVSA La Estancia is disregarding the long history of cultural production, urban infrastructure and institutions in Venezuela. According to PDVSA La Estancia's discursive constructions, Caracas would stop being an oil city once they transformed it into a socialist city. PDVSA La Estancia claimed it was making a radical shift in the relationship between the state-owned oil company and the city by discursively transcending Quintero's sterile 'culture of oil' with the mythical notion of Culture as Renewable Oil. This notion melds farming and mining language ('oil that harvests culture') to invoke culture as *the* renewable resource, the medium and ends to provide an illusion of 'renewable oil'. But this Petro-Socialist city is not a new futuristic city, it is an appropriation of the past, a re-edition of the achievements of the democratic era of the Pact of Puntofijo. The work of PDVSA La Estancia is fundamentally a large-scale restoration and appropriation of public art, buildings and urban spaces that predate the arrival of Hugo Chávez to the presidency. A 'harvest' of already existing oil wealth. PDVSA La Estancia construed a Petro-Socialist vision of a city enveloped by the Oil Social District which enables it to reframe it as a metaphorical urban oil field. Public art and public spaces become territorial markers of the oil company's appropriation of urban spaces and cultural objects, enacting a twofold colonisation of the city by oil.

Thus PDVSA La Estancia's interventions bear the 'social and cultural fruit of oil', building a direct correlation between oil and culture. PDVSA La Estancia's General Manager's vision of culture is that of the farmer-manager (Bauman, 2004) whose growing field is, in this particular case, the material space of the city reframed as an urban oil field. The advertising campaign 'We transform oil into a renewable resource for you' renders oil and culture as equivalent, as if culture is formed from sediments buried deep in the subsoil, waiting to be discovered, drilled, extracted, exploited and processed, just like crude oil. This is what is concealed behind the 'myth' (Barthes, 1993) constructed by the adverts: it allows PDVSA to expand its territorial dominion over the cities and take possession of its cultural objects as symbols of a 'future' and 'prosperous' Petro-Socialist city. The adverts in effect, negate that they were produced by a state that Hugo Chávez despised as bourgeois, capitalist and counter revolutionary. PDVSA is complicit with the Bolivarian revolution's rewriting of the past and the illusions of the New Magical State. The illusion relies on conceptualising culture as crude oil, an expedient natural resource that, unlike oil, can eternally be transformed and renewed. Oil then becomes a double-sided cultural entity: it is invoked as a singular natural force capable of producing palpable cultural effects when sown, but also a cultural product by itself. If culture is tantamount to a mineral extracted from the subsoil of the nation, then culture and the discursive and political wealth produced by Culture as Renewable Oil is, like the subsoil, the exclusive realm of the Petrostate.

That said, Culture as Renewable Oil is more than a discursive construction. By tying culture to the territory (State Space), it constructs culture as a material entity that is bound to the soil, falling under the tight control of the Petrostate. Moreover, Culture as Renewable Oil condensates the materiality oil had lost when it was reduced to the fetish of rent money (Pérez Schael, 1993; Coronil, 1997; Mitchell, 2011) becoming inextricable from land, and by that same token, from PDVSA's State Space. As an instrument of the Sowing Oil Plan, the act of sowing oil to harvest culture is an illusion of the New Magical State: oil ceases to be finite when it is sown to bear the fruits of culture. It is invoked by PDVSA La Estancia as an inexhaustible fuel for the engines of the revolution, to extend the illusions beyond the physical death of Chávez.

The illusion is completed by exclusively portraying the exuberant aspect of oil (Buell, 2014). The sanitised and impeccable giant oil worker erases any trace of pollution and dirt and conceals the catastrophic side of oil extraction by discursively and visually colluding oil and culture as a renewable resource. The giant oil workers are not interacting with cultural objects; they are portrayed manipulating the public art and public spaces as if they were heavy machinery in a refinery or an oil field. They are engaged in a mechanical process of maintenance of 'oil equipment'. For culture to find its place within the logic of Petro-Socialism, they have to be reframed as crude oil and cultural objects as oil machinery, emptied of their history and meaning.

Oil as cultural culprit

In the Venezuelan Petrostate, oil binds Bureaucratic Power and culture to the territory and its mineral-rich subsoil. This book posits that the challenge to overcome oil rentierism and oil's centrality to modern life is also cultural. Imre Zseman has aptly diagnosed that as a society 'we are incapable of imagining a world without oil' (Barrett and Worden, 2014, p. xx).

Venezuela functions as an exemplary case of a dysfunctional postcolonial Petrostate that attributes magical powers to oil, as a promise of development and wealth accumulated in the subsoil that can be transformed into material signs of modernity made manifest above ground, while failing to achieve comprehensive technological and cultural modernisation. The Petrostate inhabits the interstices between the vast oil wealth stored below ground and the signs of modernity materialised above ground. Oil is not immune to tropes of magic and utopia, it begs the question whether every Petrostate is prone to become a Magical State.

As I write this concluding chapter, Venezuela is deeply embroiled in a downward spiral of hyperinflation, economic stagnation, scarcity of goods and medicines, extreme violence and political conflict caused by the Bolivarian government's mismanagement and corruption, worsened by the steep decline in oil prices. The crash of oil prices only laid bare the severe deficiencies in state capacity that for years the influx of high oil revenues enabled the government of Hugo Chávez to mask. Yet, Venezuela does not cease to be a rentier state as political power continues to be tied to the subsoil and its mineral resources. The enduring cyclical revival of Arturo Uslar Pietri's imperative 'to sow the oil' is a palpable sign of the incapacity of the nation to imagine itself as anything other than a Petrostate. Not even Hugo Chávez, who championed an anti-capitalism and anti-neoliberalism agenda, was able to wean his ideals from oil. The concessions given by the government of Nicolás Maduro to exploit the mineral resources of the Amazon basin is a poor attempt to find an extractive alternative to oil rentierism.

Crude oil is not visible in the experience of the everyday; it is concealed and contained inside pipelines, barrels and tanks. Amid a nation that seems unable to imagine itself as anything other than a Petrostate, the giant oil worker emerged as a beacon of the oil industry to highlight the centrality of oil in contemporary life. The 'utopia of the possible' presented by the adverts of PDVSA La Estancia negates the existence of poverty, crime, scarcity and political turmoil. These illusions are untenable amid the current crisis. The utopian space of oil presented by PDVSA becomes perverse not for what it shows but by what, and who, it conceals. The adverts show an imaginary prosperity that is in stark contrast to the deterioration that is currently sweeping the country.

References

Barrett, R. and Worden, D. (eds) (2014) *Oil Culture*. Minnesota: University of Minnesota Press.

Barthes, R. (1993) *Mythologies*. London: Vintage.

Bauman, Z. (2004) 'Culture and Management', *Parallax*, 10(2), pp. 63–72.

Beblawi, H. and Luciani, G. (1987) 'Introduction', in Beblawi, H. and Luciani, G. (eds) *The Rentier State*. London: Croom Helm, pp. 1–21.

Brenner, N. and Elden, S. (2009) 'Henri Lefebvre on State, Space, Territory', *International Political Sociology*, 3(4), pp. 353–377.

Buell, F. (2014) 'A Short History of Oil Cultures; or, the Marriage between Exuberance and Catastrophe', in Barrett, R. and Worden, D. (eds) *Oil Culture*. Minnesota: University of Minnesota Press, pp. 69–88.

Coronil, F. (1997) *The Magical State: Nature, Money, and Modernity in Venezuela*. Chicago: University of Chicago Press.

Coronil, F. (2011) 'Magical History What's Left of Chavez?', in *LLILAS Conference Proceedings, Teresa Lozano Long Institute of Latin American Studies*. Latin American Network Information Center, Etext Collection.

Corrales, J. and Penfold, M. (2011) *Dragon in the Tropics. Hugo Chávez and the Political Economy of Revolution in Venezuela*. Washington D.C.: The Brookings Institution.

Karl, T. L. (1997) *The Paradox of Plenty: Oil Boom and Petro-States*. Berkeley: University of California Press.

Maass, P. (2009) *Crude World*. London: Allen Lane.

Mitchell, T. (2011) *Carbon Democracy: Political Power in the Age of Oil*. London: Verso.

Pérez Schael, M. S. (1993) *Petróleo, cultura y poder en Venezuela*. Caracas: El Nacional.

Quintero, R. (2011) 'La cultura del petróleo: ensayo sobre estilos de vida de grupos sociales de Venezuela', *Revista BCV*, pp. 15–81.

Ross, M. (2012) *The Oil Curse How Petroleum Wealth Shapes the Development of Nations*. New Jersey: Princeton University Press.

Terán Mantovani, E. (2015) 'El extractivismo en la Revolución Bolivariana: "potencia energética mundial" y resistencias eco-territoriales', *IberoAmericana*, XV(59), pp. 11–125.

Tinker Salas, M. (2009) *The Enduring Legacy: Oil, Culture, and Society in Venezuela*. Durham: Duke University Press.

Urbaneja, D. B. (2013) *La renta y el reclamo: ensayo sobre petróleo y economía política en Venezuela*. Caracas, Venezuela: Editorial Alfa.

Index

Abra Solar (art) 123
Absolute-Material Space of Caracas
 (part of the Oil Social Districts) 49, 52
Absolute Space 48; *see also* Relative
 Space; Relational Space
Acción Democrática (political party)
 27–28, 41n6
Adelman, J. 16
Agnew, John 53–54, 75
Ahearne, Jeremy 8, 10, 72, 98, 111
Alfonzo, Juan Pérez 29, 44, 96, 104, 116
Allocation State 61
Almandoz Marte, A. 25, 28
Aló Presidente (TV show) 80
Apertura Petrolera (Oil Opening
 investment programme) 33–34, 80
Armed Forces of National Liberation 86
Asamblea Nacional de La República

Banko, C. 32
Baptista, A. 3, 19–20, 26
Barrett, Ross 1, 69–70, 142
Barthes, Roland 121, 127–128, 132, 141
Bauman, Zygmunt 8, 10, 71–73, 98, 136,
 141
Beblawi, H. 2, 60–61, 141
Bennett, Tony 7, 56–57, 75, 105, 116
Betancourt, Rómulo 27–29
Bolívar, Simón 37, 80, 82–83, 85, 90–91,
 107
Bolivarian 6, 103–104; government's
 mismanagement and corruption 143;
 revolution 9, 11, 37, 41–42, 63, 65, 76,
 79, 94, 97, 103–104, 107–109, 129, 132,
 142–143; socialism 37, 103
Bolivarian Republic 35, 81, 88, 90, 139–140
Bolivariana de Venezuela 89
Brenner, Neil 7, 46, 52–55, 57, 59, 78–79,
 82–83, 85, 139

Brewer Carías, A. 27–30, 38, 89–90
British government 58–59
Buell, Frederick 71, 142
Bureaucratic Power 6–10, 39–40, 54, 57,
 59, 65, 78, 81–83, 87–88, 92, 96–103,
 105–107, 109–111, 114–120, 139; and
 authority 89; centralisation of 7, 59, 75,
 97; and culture 2, 7–8, 11, 46, 51,
 74–75, 142; disaggregation of 85, 91;
 exercise of 59, 92, 98, 139; perspective
 7, 39–40, 59, 103; and rentier state
 theory 46
Butler, Judith 72
Bye, V. 25, 27, 30

Cabrujas, José Ignacio 25, 42
Caldera, Rafael 30, 33–34
Capital District 88–89
Caracas 1, 8–9, 11–13, 16, 40–45, 49,
 88–96, 104, 108, 110, 112–120,
 123–124, 135–137, 140–141, 144;
 beautification of 112, 114–115; and the
 coalition of landowners, merchants, and
 officials 18; a petro-socialist city
 114–115, 141; ratified as the seat of
 national power 88
Caracazo revolt 5, 32
CBRV *see* Constitution of the Bolivarian
 Republic of Venezuela 1999
CDA *see* critical discourse analysis
Central Office of Coordination and
 Planning of the Presidency of the
 Republic 29, 31
Centro de Arte La Estancia 36, 97, 112;
 see also PDVSA La Estancia
Chávez, Hugo 5–6, 34–35, 37–39, 59,
 63–65, 75, 78–79, 85–87, 92–93, 96–97,
 99–101, 106–107, 138–139, 141, 143;
 and the adoption of Uslar Pietri's

Chávez, Hugo *continued*
   farming language to underpin that oil
   would be 'harvested' 105; ambitions of
   40, 138; anti-capitalist and anti-
   neoliberal rhetoric 105, 139; and the
   creation and implementation of new
   spatial strategies 9, 79, 83, 92, 105–106,
   109; government of 5, 34, 88, 139; and
   his close control over PDVSA 6; and his
   role as President 7, 10, 34–35, 59,
   81–82, 102–104, 108, 114; and his
   turning point in politics 36; nationalist
   oil policy strategies 36, 86; and Petro-
   Socialism 2, 10, 15, 30, 40, 64, 78, 97,
   101, 103; policies designed on the
   premise of inexhaustible highly priced
   oil 92, 139; Second Socialist Plan for
   the Nation 2013–2019 84–85; and the
   transfer of bureaucratic powers to
   PDVSA 78
Chevron (formally Standard Oil of
   California) 63
Colgan, J. 38–39
Comisión Presidencial para la Reforma del
   Estado 25, 32, 42
Comité de Organización Política Electoral
   Independiente 28, 33
Communal State 80, 84, 90, 103–104, 140
Congress 29–31, 33, 40
Constitution of the Bolivarian Republic of
   Venezuela 35, 38, 81, 87–89, 92
Constitutional Assembly 35
COPEI *see* Comité de Organización
   Política Electoral Independiente
COPRE *see* Comisión Presidencial para la
   Reforma del Estado
CORDIPLAN *see* Central Office of
   Coordination and Planning of the
   Presidency of the Republic
Coronil, Fernando 4–5, 9, 21, 25–27, 32,
   34, 36, 59, 70, 78, 98, 105, 134, 139,
   141–142
Corporación Venezolana de Guayana
   (Venezuelan Corporation of Guayana)
   29–30
Corporación Venezolana del Petróleo
   (Venezuelan Corporation of Petroleum)
   29–30
Corrales, Javier 6, 37, 39, 63–64, 78, 141
'counter revolutionary state' 87–88
coup d'états 32–33
Creole Oil Corporation 23–24, 130
crisis 12, 32–33, 35, 44, 69, 143;
   economic 71; energy 30–31; financial

31, 33; political 5, 32–33; social 5, 32,
   66
critical discourse analysis 10, 97–99, 101,
   116
Crouch, Colin 64
Cruz-Diez, Carlos 124
cultural studies 69; *see also* oil studies
'Culture as Renewable Oil' 7, 9, 128–129
CVG *see* Corporación Venezolana de
   Guayana (Venezuelan Corporation of
   Guayana
CVP *see* Corporación Venezolana del
   Petróleo

Darío, Rafael 117–119
Darwich, G. 5, 29–30
Delfino, M. de los Á. 87–89, 94
Dunleavy, Patrick 55–56, 76
Dunning, T. 62–63

Elden, S. 7, 11, 46–47, 52–55, 57, 59, 76,
   78–79, 82, 84–85, 93, 139, 143
energy 1, 24, 31, 35, 38, 47–48, 50, 70–71,
   86–87, 106, 109; flows of 1, 47, 50;
   industries 71; production 53, 59; world
   power 83
Engels, F. 65–66
Esfera Caracas (art) 123
Esoteric States 61
Europe 16, 19, 69–70; capitalist states in
   15; institutions of 59; societies of 80
Exoteric States 61

Fairclough, N. 98, 116
FALN *see* Fuerzas Armadas de Liberación
   Nacional (Armed Forces of National
   Liberation)
Federal District 88; *see also* Capital
   District
Federal Provinces 87
Fifth Engine of the revolution 80–81
Fifth Republic 35, 41, 88
First Engine of the revolution80
First Socialist Plan for the Economic and
   Social Development of the Nation
   2007–2013 29, 37
Five Engines of the Bolivarian Revolution
   80, 82
Five Motors of the Bolivarian Revolution
   37–38
foreign capital 4, 21, 27, 33; modernity of
   3, 22; transactions 4, 30
foreign oil corporations 3–4, 9, 21–23, 25,
   30–31, 33, 70, 109; Creole 23–24, 130;

Mobil 23, 40, 63, 97; Shell 23–24, 35, 40, 63, 130
Fossi, V. 27, 42
Foucault, Michel 56
Fourth Engine of the Bolivarian revolution 9, 79–81
Fourth National Plan 30
Fourth Republic 38, 41, 79, 81, 123, 138, 140
Fuerzas Armadas de Liberación Nacional (Armed Forces of National Liberation) 86
Functional Districts 87

Giussepe Ávalo, A.R. 34, 43, 86
Gómez, Juan Vicente 25, 27, 34
González Casas, L. 3, 22, 27
The Great Venezuela (La Gran Venezuela) 5, 30, 32, 121

Harvey, David 7, 46, 48–49, 54, 64–65, 68, 81
hydrocarbons 8, 10, 29–30, 35, 37, 49, 51, 73–75, 79–80, 88, 90, 92–93, 110–112, 135–136, 139–140

Industrial Revolution 65
Inter-American Development Bank 73

Jager, S. 98–100
Jessop, Bob 7, 55–56
Jiménez, Marcos Pérez 27–28, 34, 40, 44
Joyce, Patrick 7, 56–57, 105
Junta government 27

Karl, T.L. 2, 25–26, 28–32, 63, 141
Kozak Rovero, G. 1, 12
Kress, G. 131, 133

La Gran Venezuela (national development plan) 30
landowners 3, 21–22; apprehensions towards and rejection of oil 21; criollo 16
Latin America 9, 14–20, 23, 26–27, 31, 41, 44–45, 55, 77, 134; and the history of independent nations 18; literature and intellectual production 24; nation states 24; and Spanish colonies 15
Latin American, states 18
Law of Hydrocarbons 8, 10, 24, 49, 73–75, 88, 90, 92–93, 98, 110–112, 114, 135–136, 139–140

Lefebvre, Henri 7, 9, 46–49, 52–54, 57, 59, 65, 67–68, 82–83; categories of space 49; thinking on space 47; triad of space 46
Leibniz (associated with Relational Space) 48
Libertador Municipality 40–41, 88–89, 91
Lombardi, J. 16, 18
López, E. 85, 93
López-Maya, M. 5, 31–32, 34, 78, 105
Luciani, Giacomo 2, 11, 61–62, 75, 141
Lynch, J. 15–18

Maass, Peter 6, 33, 36, 62, 65, 105, 141
Margolis, E. 124–126, 131–132
Marín Castañeda, O. 3, 11, 22, 27, 43
Marx, Karl 6, 37, 55, 65–66, 80
Marxists 24, 36, 55, 66–67
Massey, Doreen 81
Méndez, Briceño 84–85, 94
Metropolitan District 89–91, 135, 140
Miller, Peter 7–8, 55–56, 71–72
Ministry of Energy and Mines 31, 35
Ministry of Environment and Renewable Natural Resources 31
Ministry of Infrastructure, Housing and Habitat 39
Ministry of Mines and Hydrocarbons 29
Ministry of Public Works 27
Ministry of Urban Development 31
Miranda State 89, 91
Mitchell, Tim 1, 29, 62–64, 71, 141–142
Mobil Oil Corporation (now part of ExxonMobil) 23, 40, 63, 97
Mommer, Bernard 3, 20, 37, 80
Municipal Councils 82, 113
mythology (as part of semiology) 128

National Assembly 41, 84, 88–89; and the legal foundations for the Socialist State 38; and the Permanent Commission of Oil and Energy 86
National Commission of Urbanism 27
National Constituent Assembly 86
National Constitution 104; *see also* Constitution
National Heritage Institute 123
National Historic Monument (La Estancia) 97
National Institute of Sanitary Works 27
National Project of Simón Bolívar (First Socialist Plan) 2007–2013 37, 80, 82–83, 85, 90–91, 107
New Geometry of Power 81, 85

New International Geopolitics 83
New Magical State 6, 9–10, 78–79, 81–83, 87–88, 92–93, 96, 98, 101, 105, 115, 132, 134–136, 138–139, 141–142; and Petro-Socialism 106
New National Ideal (philosophy) 28

O'Brien, D. 8, 71, 74
Ochoa Henríquez, H. 32
oil 1–6, 8–10, 12, 14–49, 52–77, 80–81, 85–87, 92–93, 96, 101–112, 114–116, 121–122, 124, 129–138, 141–144; boom times 4, 12, 30–31, 40, 43, 62, 64, 70, 76, 96, 121, 138, 141, 144; camps 3, 22–24, 30, 70, 90; and capitalism 1, 9, 23, 64–65, 69, 75, 86–87, 92, 139; concessions 30, 96; corporations 3–4, 9, 21–23, 25, 30–31, 33, 70, 109; cultural dimension of 3, 8–9, 22–23, 46, 70, 86, 133, 141; exploitation 2, 25–26, 74; exports 2, 61, 83, 85, 105; extraction 49, 91; income 107–108; industry 1–2, 4–5, 14–15, 20–21, 25, 30–31, 33, 35, 63, 70, 86, 96, 105–106, 108–110, 130–131; landowners' apprehensions towards and rejection of 21; 'magical' powers of 4, 25; production 14, 27, 31, 61; renewable oil 10; rentierism 6, 8, 33, 37–38, 40, 46, 62, 83, 86–87, 92, 103, 105–106, 109, 138–139, 141–143; self-defined revolutionary 131; social and cultural fruit of 141
oil prices 5–6, 15, 30, 32–34, 37, 39, 71, 83, 85, 96, 101, 115, 138, 143; fluctuating 63; global 31, 40, 85; high 86, 106; maintaining 33; stabilising 31
oil rent 4–7, 26, 28, 34, 36–37, 39, 60–64, 73–74, 83, 86–87, 91, 93, 106–109, 123, 132–133; distributing 26; flows of 6, 65; money 4, 14, 61; supply of 10, 106
oil reserves 21, 29, 31, 80, 103; crude 91, 115; extra-crude 80; largest crude 22, 106, 134, 136
oil revenues 3–4, 6, 11, 21, 27–31, 33, 36–37, 39–40, 61, 63, 65, 83, 105, 111; dwindling 32; generating maximum 38, 60; highest 10, 103, 106, 109, 115, 143; increasing 26, 35; potential 62; state's 2, 26, 33, 61, 105
Oil Social Districts 9–10, 49, 75, 78–79, 90–93, 96, 110, 115, 131, 135–136, 139–141
oil studies 7, 9, 69, 128–129
oil wealth 3–4, 8, 14–15, 21–26, 28, 30,

37–38, 40, 59, 63, 65, 82, 85, 139, 142; country's 31; existing 38, 138, 141; exploitation of national 3, 25, 29, 86; transient 4; vindication of 6
oil workers 1, 35, 122–137; anti-neoliberal 134; transforming oil into a 'renewable resource' 136, 142; utopian 115
OPEC 22, 29, 33, 35, 37, 61, 63, 101–104, 106, 116, 134; countries 1; inability to function as an effective cartel to stabilise oil prices 31; longevity of 105, 139; policy of maintaining oil prices by limiting production 33
Organic Law of Hydrocarbons 8, 10, 24, 49, 73–75, 88, 90, 92–93, 98, 110–112, 114, 135–136, 139–140
Organisation of Petroleum Exporting Countries *see* OPEC
Orinoco Socialist Project 80
Ortiz, Renato 19, 43
Otero, Alejandro 112, 119, 123–124

Pact of Puntofijo 5, 15, 28, 40–41, 141
Parker, D. 31, 33–36, 43, 64
Pauwels, L. 124–126, 131–132, 136
PDVSA 88, 91, 93, 96, 98, 110; enabled to absorb Caracas in the sphere of influence of 91; enabled to construct a parallel State Space through the Oil Social Districts 110; enabled to expand the Oil Social District as a dominant parallel State Space 96; enabled to extend the dominion of the Metropolitan Oil Social District across Caracas 140; enabled to instrumentalise the Organic Law of Hydrocarbons 88; enabled to interpret and implement the Organic Law of Hydrocarbons 93; enabled to interpret Article 5 of the Organic Law of Hydrocarbons as an instrument of implicit cultural policy 98; revolutionary new 90, 97, 106–107, 130, 139; State Space 10, 75, 90, 93, 99, 101, 131, 135, 140, 142
PDVSA La Estancia 8, 10, 36, 73–75, 78, 90–93, 96–98, 100, 105, 109, 111–116, 120–124, 128–136, 140–142; adverts 127–129; campaign 'we transform oil into a renewable resource for you' 49; cultural work of 49; deploys discursive constructions to perpetuate the oil rentier model 74; describes itself as 'an oasis of culture and knowledge' 97; discursive constructions claiming it is

'oil that harvests culture' in the city 132; establishes that the Oil Social Districts defined by the new Organic Law of Hydrocarbons superseded the authority of regional and municipal governments 9–10; implements the Organic Law of Hydrocarbons 75, 79; located in a Relative-Material Space 49; renders oil and culture equivalent by evoking a farming cycle 10; and the social and cultural arm of 89, 140; takes its name from the eighteenth century colonial house of Hacienda Estancia La Floresta 97

Peirce, Charles S. 10, 121, 124–126, 129, 131–132

Penfold, M. 6, 31, 37, 39, 63–64, 78, 141

Pérez, Carlos Andrés 4–5, 15, 30–34, 40, 86, 121

Pérez Schael, M.S. 2–3, 20–22, 24, 135, 141–142

Petro-Socialism 6–10, 37, 40, 75, 78–80, 82–83, 85, 92–93, 96–97, 101, 103, 105–106, 109, 115, 139; advancement of 37–38, 105–106, 109; agents of 106; and the cities 115, 141; culture 135; discourses of 49, 51; endurance of 10, 106; era of 1, 6; model of 9, 92; oil 105; and State Space 9–10, 39, 78–96, 98, 101, 105, 111, 138–140; in urban space 101

Petróleos de Venezuela Sociedad Anónima 1–2, 5–6, 8–11, 30–31, 33–39, 49, 51–52, 73–75, 78–79, 87–88, 90–93, 96–98, 104–107, 109–124, 127–143

Petrostate 2, 4–10, 21–22, 25–26, 33–34, 38, 46, 59–63, 65, 69–70, 74–75, 121–122, 134–136, 138, 142–143; centralist fiscal strategies 32; and culture in Venezuela 121; dysfunctional postcolonial 142; and its monopoly over oil wealth as a Magical State 40; and oil rents 63; Venezuelan 1–2, 4, 8–9, 15, 26, 33, 40, 57, 59, 65, 74, 105, 138, 142

Pietri, Uslar 4, 22, 37, 96, 114

Plan Siembra Petrolera (Sowing Oil Plan) 37

political parties 28, 31; Acción Democrática 27–28, 41n6; Comité de Organización Política Electoral Independiente 28, 33; Socialist Party of Venezuela 28, 37–39, 84, 106; Unión Republicana Democrática 28; United Socialist Party of Venezuela 37–38, 84, 106

PPS *see* National Project Simón Bolívar First Socialist Plan 2007–2013

Production State 61, 76

PSUV *see* United Socialist Party of Venezuela

Puntofijismo (The Pact of Puntofijo) 28, 33–35

Quintero, Rodolfo 3, 12, 22–24, 36, 44, 70, 83, 86, 95–96, 116, 133, 137, 141, 144

Quito Decree 1829 20

Ramírez, Rafael 10, 36, 38, 90, 93, 95, 97, 99–101, 106–109, 116–120, 139; a close friend of Chávez 36; condemnation of capitalism 109; president of PDVSA 38; speeches reaffirm PDVSA's allegiance to the revolution 109

Rangel, Carlos 24, 44

referendum 35, 38, 80, 88–89, 105

Relational Space 48, 50

Relative-Material Space 49

Relative Space 48–52

'renewable oil' 8–10, 52, 93, 134, 142

Representational Spaces 47–48

revolutionary Junta government 27

revolutionary PDVSA 130, 139

revolutionary State 88, 103

Rhodes, R.A.W. 57–59

Ross, Michael 62–63, 141

Sabana Grande Boulevard 123, 130, 140

Salamanca, L. 5, 32–33

San Bernardino (headquarters for Shell) 40

Sansó, Beatrice 10, 36, 93, 97, 99–100, 109–111, 114–115

Second Engine of the revolution 80–81

Second Socialist Plan for the Nation 2013–2019 85

Shadow State 57–59, 77

Shell Oil Corporation 23–24, 35, 40, 63, 130

A Short History of Oil Cultures 71

Silva-Ferrer, M. 1, 12, 34–35, 44, 74, 77

Silva Michelena, J.A. 15–16

SMP *see* State Mode of Production

socialism 6, 36–37, 49, 79–80, 84, 89, 98, 102–105, 107–110; construction of 107–108; irreversibility of 86, 106, 139; oil-based 15

Socialist Party of Venezuela 28, 37–39, 84, 106

Socialist State 6–7, 9–10, 37–38, 40, 49, 51, 78–83, 85, 87, 89, 92–93, 103–106, 109, 136, 138–140; consolidating absolute ownership and control over the territory 92; dyad oil wealth 78; new 79, 81–82, 84; and PDVSA's extension 54

Socialist State Space 9–10, 39, 78–96, 98, 101, 105, 111, 138–140

Soto, Jesús 123–124

Sowing Oil Plan (Plan Siembra Petrolera) 6, 10, 37, 96–98, 102, 105, 107, 109–112, 115, 142

Spain 2, 15–16, 43

Spanish Crown 2, 14, 20

Special Authorities 87

Special Military Regions 87

speeches 40, 86, 93, 97, 100–102, 105, 107–110, 112–113, 115, 117, 127; broadcast 81; of Chávez 102; political 101; presidential inauguration 102; presidential re-election 79; public 10, 75, 97, 100–101, 104, 106; of Rafael Ramírez 10, 100, 106–107, 109

Standard Oil of California (now Chevron) 63

Stanek, L. 47, 67

state, revolutionary 88, 103

State Mode of Production 52

state-owned oil companies 138

State Space 7–9, 46, 53–54, 59, 65, 81–82, 100–101, 136, 140, 142; administrative boundaries of 49, 78; and Bureaucratic Power 115; and Culture as Renewable Oil 99; new socialist 87, 139; PDVSA's parallel 10, 91, 136, 140; strategies 54; theorist of 7, 9, 46, 52, 54; visual metaphor of PDVSA's 10, 131

Stevenson, D. 66–68

Straka, T. 4, 19, 21–22, 30–31

Teoría Socialista sobre la Política Petrolera Venezolana (Socialist Theory of Venezuelan Oil Policy) 84–85

Terán Mantovani, E. 28, 38, 123, 138

Theory of mythical speech (Barthes) 121

Tinker Salas, M. 4, 23, 31, 35, 141

Torres, A.T. 24, 33

Unión Republicana Democrática 28

United Nations Headquarters 117

United Nations Technical Assistance Mission 28

United Socialist Party of Venezuela 37–38, 84, 106

United States *see* US

urban cultures 66, 68, 77

Urbaneja, D.B. 4–6, 26, 28, 31, 33–34, 38, 138

US 3, 23, 28, 59–60, 70, 86

Venezuela 1–6, 11–46, 62–63, 70, 76–77, 81–86, 88–90, 92–96, 103–104, 115–119, 121, 123–124, 130–131, 136–141, 143–144; and the bankrupting of the modernisation project 40, 138; contemporary 65; divided into two bodies (political and natural) 25; and the existing 'socio-territorial model' 83; and the oil industry 1–2, 4–5, 14–15, 20–21, 25, 30–31, 33, 35, 63, 70, 86, 96, 105–106, 108–110, 130–131; and the Organic Law of Hydrocarbons 8, 10, 49, 73–75, 88, 90, 92–93, 98, 110–112, 114, 135–136, 139–140; and the people of 34, 103–105; post-independence 26; and socialism 6, 36–37, 49, 79–80, 84, 89, 98, 102–105, 107–110; and society 3, 14, 21, 25–26, 31, 33–34, 40, 46, 131; transforming into a world power of oil energy 6, 38, 40, 123, 138

Wainberg, M. 31, 33–36

Walter, R. 55–56

Wolch, Jennifer 57–59

World Bank 57, 73

Yúdice, George 7–8, 10, 71–73, 75, 110, 122, 134–135

Zamora, Ezequiel 81

Zukin, Sharon 67–69

Printed and bound by CPI Group (UK) Ltd, Croydon, CR0 4YY

24/10/2024

01778282-0016